QUANTUM
DYNAMICS AND
INFORMATION

QUANTUM DYNAMICS AND INFORMATION

Editors

Robert Olkiewicz
Wrocław University, Poland

Wojciech Cegła
Wrocław University, Poland

Andrzej Frydryszak
Wrocław University, Poland

Piotr Garbaczewski
Opole University, Poland

Lech Jakóbczyk
Wrocław University, Poland

World Scientific

NEW JERSEY · LONDON · SINGAPORE · BEIJING · SHANGHAI · HONG KONG · TAIPEI · CHENNAI

Published by

World Scientific Publishing Co. Pte. Ltd.
5 Toh Tuck Link, Singapore 596224
USA office: 27 Warren Street, Suite 401-402, Hackensack, NJ 07601
UK office: 57 Shelton Street, Covent Garden, London WC2H 9HE

British Library Cataloguing-in-Publication Data
A catalogue record for this book is available from the British Library.

QUANTUM DYNAMICS AND INFORMATION
Proceedings of the 46th Karpacz Winter School of Theoretical Physics

ISBN-13 978-981-4317-43-6
ISBN-10 981-4317-43-8

Printed in Singapore.

PREFACE

The 46th Winter School of Theoretical Physics on *"Quantum Dynamics and Information: Theory and Experiment"*, organized by the University of Wrocław and the University of Opole, was held at the heart of Karkonosze Mountains in the Geovita Center in Lądek Zdrój, Poland during the period 8-13 February, 2010. The present volume contains texts of most of invited lectures delivered during the conference and a selected sample of poster presentations.

The central theme of the School was quantum dynamics, regarded mostly as the dynamics of *entanglement* and that of *decoherence*. Both these concepts appear to refer to the dynamical behavior of surprisingly fragile features of quantum systems, finite-dimensional, few and many-body structures, that are supposed to model quantum memories and stand for an arena for quantum data processing routines.

Quantum information theory is a new, dynamically evolving discipline and has many faces. Any experimental realization of a quantum information processing system is widely recognized as a difficult task because of decoherence effects, particularly dephasing, which entail loss of quantum information through randomization of the relative phase of quantum states. One of the important topics in this connection is the concept of quantum entanglement, that might possibly become the basis of a quantum computer theory and ultimate quantum computer exploitation principles.

Entanglement of quantum states is the most nonclassical manifestation of the quantum formalism. It shows up when the system consists of two (or more) subsystems and the global (pure) state cannot be written as a product of the states of individual subsystems. Then, the best knowledge of the whole of a composite system may not include nor reflect a complete knowledge of its subsystems. This notion can be generalized to mixed states, and a mixed state of a quantum system is entangled if the corresponding density matrix cannot be expressed as a convex combination of tensor products of density matrices of its subsystems.

A fundamental nonclassical aspect of entanglement was recognized already in 1935 by Schrödinger who was led to say that it is "not one but rather the characteristic trait of quantum mechanics". In the present-day the theory of entanglement is well established and known to play a central role in the quantum information science, e.g. quantum communication, quantum cryptography and quantum computing.

The entanglement property is not often considered to be a physically (experimentally) accessible property of a quantum system. However, it may stand for a resource that can help in tasks such as the reduction of classical communication complexity, frequency standards improvements and clock synchronization. From the theoretical point of view, the entanglement concept has a very complex structure and the theory of entanglement tries to give answers to a number of questions of primary importance: how to optimally detect the entanglement, how to protect enanglement against degradation (fragility issue), and how to characterize it, control and quantify.

In the context of quantum dynamics, the most important aspect of entanglement is that it is fragile with respect to any noise resulting from an interaction with the environment. This may lead to the dissipation and destruction of correlations and due to that, entanglement may disappear even though it was initially present in the quantum system state. To control a corresponding process of disentanglement, it is important to understand details of the complex nature of the evolution of entanglement in open quantum systems.

In most models of quantum open systems, quantum coherence may be destroyed asymptotically in time, but in some cases entanglement still may abruptly and completely disappear (die out) in finite time. This phenomenon, named *entanglement sudden death* (ESD), has been explored in a variety of contexts: theoretically and experimentally in continuous and discrete systems, few body and many-body systems. Physical systems that have been examined include electrons on the solid state lattice subject to an electromagnetic field, photons in a system of mirrors and beam splitters, atoms confined in cavities. The omnipresent noise can be both classical and quantum in origin.

The entanglement of qubits, which is at the heart of promising speed-up offered by quantum computations, usually enhances the effects of decoherence, and thus causes a faster decay of computational fidelity. Therefore, the study of decoherence as a phenomenon which leads to a corruption of the information stored in the system as well as errors in computational steps is strongly intertwined with the field of quantum information.

The first challenge one has to deal with is the role of quantum correlations and entanglement in the process of decoherence and the description of the robustness of some entangled states under typical environmentally induced decoherence phenomena. The other key problems include engineering ways to modify and eliminate decoherence in applications of quantum information processing by exploiting the so-called decoherence free subspaces, entanglement distillation or dynamical decoupling procedures. On the other hand, contribution to experiments on decoherence to further understand the quantum to classical transition to read out results of quantum computations should take place.

Finally, it should be pointed out that the existing protocols for dynamical dephasing control during all stages of quantum information processing, namely the storage and gate operations, cannot be readily implemented in available experimental set-ups, such as ion traps, quantum dots or optical lattices, and so they require further investigations on both the experimental and theoretical level.

Spontaneous emission in the two-atomic system is an example of the noise which diminishes entanglement and may lead to the entanglement sudden death. On the other hand, the process of photon exchange can induce correlations between atoms which can partially overcome decoherence. As a result, some amount of entanglement present in the system can survive. Moreover, there is a possibility that this process can entangle separable states of two atoms.

Such *dynamical creation of entanglement* in the presence of a noisy environment, have attracted a great deal of attention. In particular, the production of robust asymptotic entanglement for closely separated atoms as well as the existence of transient entanglement in the case of arbitrary separations have been established. In that case, the dynamics of entanglement is so complex that the phenomena of revivals of entanglement that has already been destroyed or even *delayed sudden birth of entanglement* can occur. For general Markovian completely positive dynamics of two atoms immersed in a common bath, dynamical creation of entanglement has been established.

All that is merely a small part of the whole entanglement, decoherence and information intertwine. There exist excellent reviews on some of those topics: quantum entanglement as an information resource, entanglement in many-body systems, information measures and thermodynamically justified processing of information. Grand reviews were published as well on various aspects of decoherence and its possible role for the classical-quantum intertwine.

Most of the related experimental topics that include: single trapped atom manipulations, single atom-photon interaction in a cavity (cavity quantum electrodynamics), advances in large molecules interferometry, interferometry of Bose–Einstein condensation and a multitude of experiments on quantum information transfer (teleportation issue included), have nowadays received ample publications coverage.

That gave us, Editors for this volume, some freedom in slightly reshuffling the accents. While editing the present volume we have intentionally pushed its contributors to present topics that still remain insufficiently explored, albeit being of an utmost importance.

The programme of the School has been shaped with the help of the scientific committee comprising: V.M. Akulin, Ph. Blanchard, S. Chwirot, R. Horodecki, P. Knight, J. Rembieliński, R.F. Werner, A. Zeilinger, P. Zoller and M. Żukowski. We convey our thanks to all of them.

The School has been financially supported by the University of Wrocław, University of Opole, Polish Academy of Sciences and the European Physical Society.

The Editors
July 2010

CONTENTS

Quantum memories and Landauer's principle

Robert Alicki

Instytut Fizyki Teoretycznej i Astrofizyki
Uniwersytet Gdański
ul. Wita Stwosza 57, 80-952 Gdańsk, Poland
e-mail: fizra@univ.gda.pl

Two types of arguments concerning (im)possibility of constructing a scalable, exponentially stable quantum memory equipped with Hamiltonian controls are discussed. The first type concerns ergodic properties of open Kitaev models which are considered as promising candidates for such memories. It is shown that, although the 4D Kitaev model provides stable qubit observables, the Hamiltonian control is not possible. The thermodynamical approach leads to the new proposal of the revised version of Landauer's principle and suggests that the existence of quantum memory implies the existence of the perpetuum mobile of the second kind. Finally, a discussion of the stability property of information and its implications is presented.

Keywords: quantum computing, quantum memory, Landauer's principle, laws of thermodynamics.

1. Introduction

The most challenging idea in *Quantum Information* is the possibility of *fault-tolerant quantum information processing* which violates two rather fundamental principles.

The first one can be called **(Classical) Complexity–Theoretic Church–Turing Thesis** (CTT), which states[a]:
A probabilistic Turing machine can efficiently simulate any realistic model of computation.

The second principle is **Bohr's Correspondence Principle** (BCP) expressed in a form:
Classical physics and quantum physics give the same answer when the systems become large,

---

[a]This formulation is not due to Church or Turing, but rather was gradually developed in complexity theory. "Efficiently" means up to polynomial-time reductions.

or in other words:

For large systems the experimental data are consistent with classical probabilistic models.

Indeed, the Shor's algorithm, when realized in a fault-tolerant way, allows factoring of numbers in a polynomial time. This task is believed to be unfeasible using classical probabilistic computers and hence CTT must be violated. On the other hand quantum computing should be scalable and therefore according to BCP at a certain scale a quantum computer should be described by a classical probabilistic model.

Although there exists a well-developed theory of fault-tolerant quantum computation (FTQC) based on *error correction* schemes,[1–4] its phenomenological assumptions are doubtful and in author's opinion not convincing.[5–8] The basic idea of error correction is the following. One assumes that the (mixed) states of, say, single logical qubit can be identified with a subset S_q of the density matrices for a larger, but still controlled system and T is a completely positive trace preserving map (CP-map) modeling errors (noise) caused by the interaction with an uncontrolled environment. The *error correcting map* is another CP-map R such that

$$RT\rho_q = \rho_q, \text{ for any } \rho_q \in S_q. \tag{1}$$

As any CP-map W is a contraction with respect to the trace norm i.e.

$$\|W\rho - W\rho'\|_1 \leq \|\rho - \rho'\|_1, \text{ for any } \rho, \rho' \tag{2}$$

we have

$$\|\rho_q - \rho'_q\|_1 = \|RT\rho_q - RT\rho'_q\|_1 \leq \|T\rho_q - T\rho'_q\|_1 \leq \|\rho_q - \rho'_q\|_1 \tag{3}$$

what implies $\|T\rho_q - T\rho'_q\|_1 = \|\rho_q - \rho'_q\|_1$ and hence the existence of a unitary operator U such that

$$T\rho_q = U\rho_q U^\dagger \text{ for any } \rho_q \in S_q. \tag{4}$$

The equality (4) means that the degrees of freedom which represent the logical qubit are not essentially affected by the noise and therefore one should rather speak about *error avoiding* than *error correcting* schemes. The same argument is valid also for classical systems where U is a Koopman's unitary map induced by a measure preserving transformation on the phase-space. The above conclusion is fundamental for the feasibility of FTQC which is equivalent to the existence of quantum systems - *quantum memories* - which possess quantum subsystems sufficiently stable with respect to noise.

To be more precise, one can define a *quantum memory* for a single *encoded qubit* as the quantum system interacting with a heat bath at the temperature T which consists of N microscopic subsystems (e.g. spins-$1/2$, *physical qubits*) such that a certain subalgebra of observables isomorphic to the algebra M_2 of a qubit is *exponentially stable* for the temperatures low enough. Exponential stability means that the relaxation times of the encoded qubit observables increase exponentially with N and it is a necessary condition to perform computations of an arbitrary, polynomial in N length.

A quantum memory admits *Hamiltonian control* if one is able to implement the Hamiltonians of the form

$$H_{con}(t) = X \otimes F_x(t) + Y \otimes F_y(t) + Z \otimes F_z(t) \tag{5}$$

where X, Y, Z are standard hermitian operators which span the algebra M_2 (analogs of Pauli matrices) and F_x, F_y, F_z are time-dependent hermitian operators of a generic quantum system or, as a limit case, external classical fields.

It seems that only an exponentially stable quantum memory with Hamiltonian control can be useful for fault-tolerant quantum information processing.

In the next Section the fundamental mechanism of stability for classical information is briefly discussed. Then, in the following Sections two approaches to the problem of existence of quantum memory are presented. The first, a constructive one, involves designing of quantum N spin-$1/2$ systems with particular Hamiltonians which are supposed to generate stable encoded qubit observables when weakly coupled to a heat bath at temperatures below a certain critical one. In particular, the results for 2D and 4D Kitaev models are discussed which show that while the later model contains exponentially stable qubit observables, their Hamiltonian control seems to be not feasible. The second approach is based on thermodynamics. It is argued that the information processing based on stable information carriers leads to reexamination of the Landauer's principle. From this new formulation it follows that a quantum memory with a Hamiltonian control cannot be realized in Nature, because its existence implies the existence of a *perpetuum mobile of the second kind*. This raises questions about the very nature of the notion of information, both in the classical and quantum context.

2. Stable classical memories

The macroscopic world around us is full of bodies with well-defined positions and shapes and therefore able to encode information. Even much smaller objects, like macromolecules, exist in stable configurations which carry, for example, biologically relevant information. The mechanisms which provides stability of such "memories" with respect to thermal noise is universal. Matter occupies rather local minima of free energy than a global minimum corresponding to a single uniform state. Those minima are separated by free energy barriers F_N which are proportional to the systems sizes $F_N \sim N$. The Boltzmann factor $\exp -F_N/kT$ dominating the thermal transition rates between the minima yields life-times of such metastable states growing exponentially with N. Even for relatively small molecules these life-times became much longer than the time scale of cosmological processes in the Universe.

As a simple example one can briefly discuss the classical Ising models which are able to encode a single bit in terms of the magnetization sign. Compare the mean-field Ising model ($\sigma_j = \pm 1$) with the Hamiltonian

$$H_N^{mf} = -\frac{J}{2N} \sum_{i,j=1}^{N} \sigma_i \sigma_j \tag{6}$$

with the 1D-Ising Hamiltonian

$$H_N^{1D} = -J \sum_{j=1}^{N} \sigma_j \sigma_{j+1} \tag{7}$$

The energy difference between two configurations $+++++++++++++$ and $+++\underbrace{----}_{k-times}+++++$ is given by

$$\Delta E^{mf} = Jk + \frac{k^2}{2N} \; , \; \Delta E^{1D} = 2J \tag{8}$$

and shows the mechanism of bit's protection against noise. While for the 1D Ising model the energy difference does not depend on the number k of flipped spins, for the mean-field model it grows with k (the same holds for 2D, 3D,...). At finite temperatures one has to take into account the entropy contribution to the free energy which suppress the effect of protection at high temperatures leading to the ferromagnetic phase transition phenomenon for 2D, 3D,.., and mean-field Ising models. The question arises: *Does a similar mechanism can protect encoded qubit?*

3. Kitaev models

The Kitaev models in $D = 2, 3, 4$ dimensions[9,10] are spin-1/2 models defined on a D-dimensional lattice with a toric topology. The Hamiltonian always possess the special structure:

$$H = -\sum_s X_s - \sum_c Z_c, \tag{9}$$

where, $X_s = \otimes_{j \in s} \sigma_j^x$, $Z_c = \otimes_{j \in c} \sigma_j^z$ are products of Pauli matrices belonging to certain finite sets on the lattice, called *stars* and *cubes*. They are chosen in such a way that all X_s, Z_c commute forming an abelian subalgebra $\mathcal{A}_{ab}$ in the total algebra of $2^N \times 2^N$ matrices. The noncommutative commutant of $\mathcal{A}_{ab}$, denoted by $\mathcal{C}$, is a natural candidate for the subalgebra containing encoded qubit observables.

The stability with respect to thermal noise has been studied within Markovian models with semigroup generators derived by means of the Davies weak coupling procedure. The obtained Markovian master equations possess all properties necessary from the phenomenological point of view: any initial state relaxes to the Gibbs thermal equilibrium, detailed balance property holds.[11] For Kitaev models which are *ultralocal* such derivations are mathematically sound and simple. This makes the analysis of spectral properties of the Davies generators feasible, but still too involved to be reproduced here; we refer the reader to[12-14] for details.

For the 2D Kitaev model the spectrum of the Hamiltonian (9) is particularly simple. The ground state is four-fold degenerated and any excited state is fully characterized by one of the ground states and the positions on the lattice occupied by single excitations called *anyons*. There are two types of anyons, corresponding to stars (X-type) or cubes (Z-type) and their total numbers are even. The relative energy of such a state with respect to ground states is equal to the total number of anyons.

To describe thermal relaxation, one couples all spins to individual identical heat baths by Hamiltonian terms containing σ_j^x, σ_j^z. Then, the form of the Markovian master equation for the density matrix in the interaction picture, and with a uniform normalization of relaxation rates, is the following

$$\frac{d\rho}{dt} = \frac{1}{2} \sum_{j=1}^{N} \left\{ \left([a_j, \rho\, a_j^\dagger] + [a_j\, \rho, a_j^\dagger] \right. \right.$$

$$\left. +e^{-2\beta} \left([a_j^\dagger, \rho\, a_j] + [a_j^\dagger\, \rho, a_j] \right) \right) - [a_j^0, [a_j^0, \rho]] \right\}$$

$$+\frac{1}{2} \sum_{j=1}^{N} \left\{ \left([b_j, \rho\, b_j^\dagger] + [b_j\, \rho, b_j^\dagger] \right. \right.$$

$$\left. +e^{-2\beta} \left([b_j^\dagger, \rho\, b_j] + [b_j^\dagger\, \rho, b_j] \right) \right) - [b_j^0, [b_j^0, \rho]] \right\}. \tag{10}$$

Instead of defining here the operators a_j, a_j^0, b_j, b_j^0 their physical interpretation is outlined. The operator a_j $(a_j^\dagger)$ annihilates (creates) a pair of type-Z anyons attached to the site j , while a_j^0 generates diffusion of anyons of the same type. Similarly, the operators $b_j, b_j^\dagger, b_j^0$ correspond to the type-X anyons.

The structure of the Hamiltonian (9) and the master equation (10) imply that the $2D$-Kitaev model is equivalent to a gas of noninteracting particles which are created/annihilated in pairs and diffuse. Heuristically, no mechanism of macroscopic free energy barrier between different phases is present which could be used to protect even a classical information. Mathematically, it was proved that the dissipative part of the Davies generator (in the Heisenberg picture) possesses a spectral gap independent of the size N and therefore no metastable observables exist in this system.

In contrast to the 2D case the 4D Kitaev model can be described by a picture similar to droplets in the 2D-Ising model.[10] The excitations of the system are represented by closed loops with energy proportional to the loops' length providing the mechanism of a macroscopic energy barrier separating topologically nonequivalent spin configurations. The 3D model is an intermediate case, this mechanism works for the one type of excitations only. Therefore, only encoded "bit" is protected but not the "phase". The structure of the evolution equation, for all dimensions is always similar to (10) with the operators $a_j^\dagger, b_j^\dagger$ creating excitations of two types and a_j^0, b_j^0 changing the shape of excitations but not their energy.

The rigorous arguments concerning 4D case are presented in the paper.[14] Generally, for all $D = 2, 3, 4$ one can define *bare qubit observables* $X^\mu, Z^\mu \in \mathcal{C}$ where $\mu = 1, ..., D$. They are products of the corresponding Pauli matrices over topologically nontrivial loops (surfaces) which are not unique. However, only for 4D case there exist exponentially metastable *dressed qubit observables* $\tilde{X}^\mu, \tilde{Z}^\mu \in \mathcal{C}$ with $\mu = 1, 2, 3, 4$ related to the bare ones by the formulas

$$\tilde{X}^\mu = X^\mu F_x^\mu, \quad \tilde{Z}^\mu = Z^\mu F_z^\mu. \tag{11}$$

where F_z^μ, F_x^μ are hermitian elements of the algebra $\mathcal{A}_{ab}$ with eigenvalues ± 1. On the other hand, bare qubit observables are highly unstable with relaxation times $\sim \sqrt{N}$. The metastability of (11) is proved using the Peierls argument applied to classical "submodels" of the 4D-Kitaev model generated either by $- \sum_s X_s$ or $- \sum_c Z_c$ and holds below certain critical temperature.

Any metastable observable (say $\tilde{X}^\mu$) is given by the following procedure, which operationally determines its outcomes:

1. Perform a measurement of all observables σ_j^x.

2. Compute the value of X^μ (multiply previous outcomes for spins belonging to the "surface" which defines X^μ).

3. Perform a certain classical algorithm (polynomial in N) which allows to compute from the σ_j^x- measurement data the value ± 1 of "correction", i.e. the eigenvalue of F_x^μ, and multiply it by the bare value to get the outcome of $\tilde{X}^\mu$.

The procedure of above provides the structure of 4-qubit subalgebra which is exponentially stable with respect to thermal noise below the critical temperature. However, the dressed qubit observables are given in terms of a certain algorithm involving a destructive and not repeatable measurement on individual spins. Those qubit observables cannot be used to construct Hamiltonians and therefore this type of memory is not equipped with Hamiltonian controls. In the author's opinion this fact reflects fundamental difficulties with the implementation of the idea of FTQC. This leads to a natural question: *Can phenomenological thermodynamics provide restrictions or even no-go theorems for Quantum Memory, or generally for Quantum Information Processing?*

4. Thermodynamics of information processing

There exists a deep similarity between the complexity theory and thermodynamics. In both cases the predictions have asymptotic character with respect to the parameter N which in the first case denotes the size of the input and in the second case the number of elementary physical constituents (atoms, spins, photons,...), called *particles*. To make these relations closer one needs to reformulate the laws of thermodynamics which are usually expressed in a natural language and have a common sense character.

In[15] the following reformulation has been proposed:

Zero-th Law:

Any N-particle system coupled to a thermal bath relaxes to the (possibly nonunique) thermodynamical equilibrium state at the bath's temperature with relaxation time growing at most polynomially in N.

Second Law:

It is impossible to obtain an effective process such that the unique effect is the subtraction of a positive heat from a reservoir and the production of a positive work of the order of at least $k_B T$. By effective process we mean a process which takes at most polynomial time in the number of particles N.

Fluctuation Theorem
For a system consisting of N particles the probability of observing during time t an entropy production opposite to that dictated by the second law of thermodynamics decreases exponentially with Nt.

One should notice that the Zero-th Law provides now a proper definition of the thermal equilibrium which includes multiple thermodynamical phases and metastable configurations corresponding to local minima of the free energy. Fluctuation Theorem[16] determines the limits of applicability of the phenomenological thermodynamics.

4.1. *Landauer's Principle*

The direct connection between thermodynamics and information theory is given by the Landauer's Principle[17] in the formulation of Bennett:[18]
Any logically irreversible manipulation of information, such as the erasure of a bit or the merging of two computation paths, must be accompanied by a corresponding entropy increase in non-information bearing degrees of freedom of the information processing apparatus or its environment.
Specifically, each bit of lost information will lead to the release of an amount $k_B T \ln 2$ of heat, where k_B is the Boltzmann constant and T is the absolute temperature of the circuit.

There exist two kind of "proofs" of the Landauer's principle. The first, and correct one is essentially based on Szilard's discussion of the *Maxwell demon*. A Szilard engine consists of a single particle in a box coupled to a heat bath. Knowing which half of the box is occupied by a particle one can close a piston unopposed into the empty half of the box, and then extract $k_B T \ln 2$ of useful work using isothermal expansion to its original equilibrium state. Assuming that the Second Law holds the "measurement process" needs at least the same amount of work which is attributed by Bennett to a single bit erasure in the process of reseting of the measuring device.

One should notice that the argument of above does not apply to the situation where a bit of information is encoded in two possible equilibrium (metastable) states. Namely, after a measurement one cannot use the relaxation process, like isothermal expansion, to extract work and leave the system in a completely mixed state, because such relaxation needs exponentially long times.

The other type of "proof", which is incorrect, employs the total entropy balance for the information bearing subsystem plus the environment. In its simplest (and quantum) version the process of a bit's erasure is described by the map

$$\rho_{in} \otimes \omega_{in} \Longrightarrow \rho_{out} \otimes \omega_{out}, \quad \rho_{in} = \frac{1}{2}(|0\rangle\langle 0| + |1\rangle\langle 1|), \quad \rho_{out} = |0\rangle\langle 0|. \quad (12)$$

Here, the state $|0\rangle$ is a fixed reference state of the information bearing system and the initial unknown state ρ_{in} is completely mixed. The density matrices ω_{in} and ω_{out} describe the initial and final states of the heat bath. Assuming the validity of the Second Law for the total isolated system one obtains the estimation for the entropy balance ($S(\rho) = -k_B \text{Tr}\rho \ln \rho$)

$$S(\omega_{out}) - S(\omega_{in}) \geq k_B \ln 2 \quad (13)$$

what implies that at least $k_B T \ln 2$ of heat is dissipated into the heat bath during the reseting process. The weak point of this argument is related to "discontinuity" of entropy for large systems expressed in terms of Fannes inequality[19] for two close density matrices of a system with D-dimensional Hilbert space and $\| \cdot \|_1$ denoting the trace norm

$$|S(\rho) - S(\rho')| \leq \|\rho - \rho'\|_1 \ln D - \|\rho - \rho'\|_1 \ln(\|\rho - \rho'\|_1) . \quad (14)$$

Due to (14) even an infinitesimally small perturbation of the initial product state in (12), which is always present, can cause a substantial change in the total entropy.

Another important case illustrating this problem is the Markovian dynamics of an open system. Here the state of the total system is well-approximated by the product $\rho(t) \otimes \omega_B$ where $\rho(t)$ is a solution of the Markovian master equation and ω_B is a fixed equilibrium state of a bath. Obviously, this product form is not consistent with the constant entropy of the total Hamiltonian system. The missing entropy is hidden in the small correction terms describing the residual system - bath correlations and small local perturbations of the bath's state.

Summarizing, one can trust only the argument proposed by Szilard which is entirely based on the Second Law and which leads to the following:

Revised Landauer's Principle for Measurement

A measurement which allows to distinguish between two different states of a system coupled to a heat bath:

a) needs at least $k_B T \ln 2$ of work if those states relax to their uniform statistical mixture,

b) does not need a net amount of work if those states are equilibrium ones.

4.2. Quantum memory as perpetuum mobile

Following[15] one can construct a model of a *perpetuum mobile of the second kind* under the assumption that one possesses a single-qubit exponentially stable quantum memory with a Hamiltonian control. The device contains besides the memory a quantum version of the Szilard engine. It is a two-level system with controlled energies of both levels. The system relaxes to the Gibbs state due to the interaction with a heat bath. Assume that the initial state is the one of those eigenstates and the initial energies are both equal to zero. If one knows which level is occupied then one can quickly increase the energy of the second level to the value $E >> k_B T$. In the next step one couples the system to a heat bath and slowly decreases the energy of the second level back to zero. It is easily to compute that during this process an amount of work $W \simeq k_B T \ln 2$ is subtracted from the heat bath.[20] In the standard setting one concludes that the process of acquiring information about the initial state, in a cyclic process, needs at least the same amount of work. Here, the quantum memory is used to prepare a given initial state. Namely, according to the revised Landauer's principle a measurement of a state of quantum memory in a given basis does not cost work. Knowing the state of memory one can swap it with the relaxing system and then start the process of extracting work. The swap operation is unitary and hence does not cost work and can be realized using the Heisenberg-type interaction Hamiltonian[21] between the memory and the relaxing qubit

$$H_{\text{int}}(t) = \lambda(t)\big(X \otimes \sigma^x + Y \otimes \sigma^y + Z \otimes \sigma^z\big). \tag{15}$$

Here σ^k are Pauli matrices for the relaxing qubit and $\lambda(t)$ is a time-dependent coupling constant. The construction of such a Hamiltonian is guaranteed by the assumption of Hamiltonian control for the quantum memory. Finally, the net effect of the whole cyclic process is the subtraction of heat from the bath and the production of work what violates the Second Law of Thermodynamics.

5. Two types of information?

The arguments of the previous Section can be also applied to a classical stable memory. The revised Landauer's Principle is valid in the classical case also and it is not difficult to give an example of a classical Hamiltonian swap operation.[b] To avoid the conflict with the Second Law one has to conclude

---

[b]It can be done for two classical systems with a single degree of freedom each and the Hamiltonian quadratic in position and momenta.

that the Hamiltonian (reversible) gates are incompatible with stability of encoded information. Indeed, taking as a model of classical one-bit memory a particle in a double-well potential, it is rather obvious that friction is a necessary ingredient to stabilize information. Applying a Hamiltonian gate by kicking a particle from one well to the other one never reaches a final stable state but rather oscillations between two values of bits as long as friction is not at work. Obviously, in the existing computers all gates are strongly irreversible.

It seems that stability of information is an important ingredient of its very definition. One often does not make difference between "pseudo-information" encoded e.g. in temporal positions of gas particles and "information" encoded e.g. in a shape of a macroscopic body. The previous discussion on the Landauer's principle and the intuitive analysis of simple examples allow to characterize the fundamental features of both notions:

1) **Pseudo-Information** is unstable with respect to thermal noise; can be used to extract work from a heat bath and hence its acquiring costs work at least $k_B T \times (Shannon\ entropy)$; can be processed reversibly.

2) **Information** is stable with respect to thermal noise (at least below a certain temperature, with life-time scaling $\sim \exp(\gamma N)$); its acquiring does not cost work; its Shannon entropy has no thermodynamical meaning; must be processed irreversibly with a cost of a single gate $\sim \gamma N k_B T$ of work.

The intrinsic lack of stability with respect to thermal noise makes the practical applicability of pseudo-information very limited. In the author's opinion Quantum Information is, unfortunately, an example of pseudo-information.

Acknowledgments

The author thanks Michał Horodecki for discussions. This work is supported by the Polish Ministry of Science and Higher Education grant PB/2082/B/H03/2010/38.

Bibliography

1. Knill E, Laflamme R, and Żurek W H 1998 Resilient quantum computation *Science* **279** 342–345
2. Aharonov D and Ben-Or M 1997 Fault-tolerant quantum computation with constant error rate in *Proc. of the 29th Annual ACM Symposium on Theory of Computing (STOC) 1997* 176–188 `quant-ph/9910081`
3. Gottesman D 2000 Fault-tolerant quantum computation with local gates *J. Mod. Opt.* **47** 333–345 `quant-ph/9903099`

4. Preskill J 1998 Reliable quantum computers *Proc. Roy. Soc. Lond. A* **454** 385–410 `quant-ph/9705031`
5. Alicki R, Horodecki M, Horodecki P, and Horodecki R 2001 Dynamical description of quantum computing: Generic nonlocality of quantum noise *Phys. Rev. A* **65** 062101 `quant-ph/0105115`
6. Alicki R, 2006 Quantum error correction fails for Hamiltonian models *Fluctuation and Noise Letters* **6** c23
7. Alicki R, Lidar D, and Zanardi P 2006 Internal consistency of fault-tolerant quantum error correction in light of rigorous derivations of the quantum markovian limit *Phys. Rev. A* **73** 052311 `quant-ph/0506201`
8. Dyakonov M I 2006 Is Fault-Tolerant Quantum Computation Really Possible? `quant-ph/0610117`
9. Kitaev A Y 2003 Fault-tolerant quantum computation by anyons *Annals Phys.* **303** 2–30
10. Dennis E, Kitaev A, Landahl A, and Preskill J 2002 Topological quantum memory *J. Math. Phys.* **43** 4452–4505 `quant-ph/0110143`
11. Alicki R, and Lendi K 2007 *Quantum Dynamical Semigroups and Applications* II-nd edition, LNP 717, Springer, Berlin
12. Alicki R, Fannes M and Horodecki M 2007 A statistical mechanics view on Kitaev's proposal for quantum memories, *J. Phys. A: Math. Theor.* **40** 6451-6467 `arXiv:quant-ph/0702102`
13. Alicki R, Fannes M and Horodecki M 2009 On thermalization in Kitaev's 2D model *J. Phys. A: Math. Theor.* **42** 065303(18pp); `arXiv:0810.4584`
14. Alicki R, Horodecki M, Horodecki P and Horodecki R 2009 On thermal stability of topological qubit in Kitaev's 4D model *Open Systems and Information Dynamics* (in print) `arXiv:0811.0033`
15. Alicki R Quantum memory as a perpetuum mobile of the second kind `arxiv:0901.0811v4`
16. Evans D J and Searles D J 2002 The Fluctuation Theorem *Advances in Physics* **51** 1529 - 1585
17. Landauer R 1961 Irreversibility and heat generation in the computing process *IBM Journal of Research and Development* **5** 183-191
18. Bennett C H 2003 Notes on Landauer's principle, Reversible Computation and Maxwell's Demon *Studies in History and Philosophy of Modern Physics* **34** 501-510
19. Fannes M 1973 A continuity property of the entropy density for spin lattice systems *Commun. Math. Phys.* **31** 291-294
20. Alicki R, Horodecki M, Horodecki P and Horodecki R 2004 Thermodynamics of Quantum Information Systems – Hamiltonian Description *Open Systems and Information Dynamics* **11** 205-217
21. Loss D and DiVincenzo D P 1998 Quantum Computation with Quantum Dots *Phys. Rev. A* **57**, 120-126

Asymptotic entanglement in open quantum systems

F. Benatti

Department of Physics, University of Trieste,
Trieste, Italy
E-mail: benatti@ts.infn.it

We consider the behavior of entanglement in open quantum systems and show that, despite decoherence and in contrast to expectations, suitably engineered couplings to the environment may generate quantum correlations in unexpected ways.

Keywords: open quantum systems; dissipative generation of entanglement; entanglement production.

1. Introduction

In recent years, the development of quantum information theory has completely transformed the perception of entanglement; the status of such a notion indeed changed from that of an epistemological riddle relative to the completeness of quantum mechanics to a physical resource that allows performing informational tasks, namely manipulation and transmission of information, that would be impossible by using classical resources.[1-3] The creation of entangled states robust against decoherence is a main concern of quantum information theory, a major source of decoherence being non–negligible couplings to the environment in which quantum systems are immersed.

When the couplings to the environment are weak enough, one deals with open quantum systems for which various techniques allow one to derive a reduced dynamics which is free of memory effects, consists of a semigroup of completely positive maps and is fully characterized by a master equation of Lindblad type.[4,5] The noisy and dissipative effects described by such master equations generically result in decoherence and in depletion of entanglement; however, it turns out that if, suitably engineered, the coupling to the environment can be used to generate entanglement and even to make it persist asymptotically.[6-13]

After briefly reviewing the salient facts of open quantum system dynamics and of entanglement, the focus will be upon the generation of entanglement in open quantum systems consisting of two qubits and upon its behavior on short and large times. In order to better emphasize the possibility of generating entanglement by dissipation, we will tackle a conjecture recently put forward [14] stating that the production of entropy will always last longer than that of entanglement. We show that the conjecture is violated when there is asymptotic entanglement.

In the second part of the contribution, we will consider three equal qubits and illustrate the possibilities set in store by higher dimensionality via discussing a protocol that consists in appending a third qubit to a pair of two qubits, switching on a highly symmetric coupling to an environment, letting equilibrium to be reached and finally eliminating one qubit. We show that the remaining two qubits can get more asymptotic entanglement than what they would by direct immersion in the same environment.

2. Quantum systems in their environment

In the following we shall focus upon finite level systems S consisting of two bipartite systems weakly interacting with their environment E. The observables of the system of S will be Hermitean matrices $M_d(\mathbb{C})$ and its states either by vectors $|\psi\rangle \in \mathbb{C}^d$, if pure, or by density matrices, that is by positive semi-definite matrices $\rho \in M_d(\mathbb{C})$ of trace 1.

Firstly, we briefly resume how a memoryless master equation can be obtained for an open quantum system, that is how a Markovian, dissipative time-evolution, known as *quantum dynamical semigroup*, can be extracted from the global reversible time-evolution of the compound system $S + E$. More details can be found in standard reference books [4,5]

Weak-coupling limit The compound system $S + E$ is treated as closed so that its reversible dynamics is generated by a total Hamiltonian

$$H_T = H_S + H_E + \lambda H_I , \qquad (1)$$

where $H_{S,E}$ are the Hamiltonian operators of system and environment, while

$$H_I = \sum_\alpha X_\alpha^S \otimes X_\alpha^E \qquad (2)$$

describes the interaction among them by means of operators $X_\alpha^{S,E}$ of the system, respectively of the environment. The weakness of the coupling is measured by an a–dimensional coupling constant $\lambda \ll 1$.

If the unitary time-evolution: $U_t = \exp(i\,t\,H_T)$ $(\hbar = 1)$ acts on a factor-ized initial states

$$\rho_T = \rho_S \otimes \rho_E \,, \tag{3}$$

by eliminating the environment degrees of freedom via a partial trace over the environment Hilbert space, one extracts the time-evolution of S, $\rho_S \mapsto \rho_S(t)$ as follows

$$\rho_S \mapsto \rho_S(t) = \mathrm{Tr}_E\Big(U_t\,\rho_S \otimes \rho_E\,U_t^\dagger\Big) \tag{4}$$

$$\partial_t \rho_S(t) = -i\,[H_S\,,\,\rho_S(t)] + \lambda^2 \int_0^t du\,\mathbb{D}_u[\rho_S(t-u)] \tag{5}$$

The maps $\Lambda_t : \rho_S \mapsto \rho_S(t)$ are trace-preserving and completely positive; that is, the map $\Lambda_t \otimes \mathrm{id}_n$ preserves the positivity of all states of the com-pound system $S + S_n$ where S_n is any n-dimensional ancilla.

Having traced out part of the total system, irreversibility sets in and, because of the memory effects embodied by the kernel of the integro–differential equation (4), the maps Λ_t do not constitute a forward–in–time semigroup. However, the environmental effects can only be visible on the slow time-scale $\tau = \lambda^2\,t$, where the memory effects can be neglected and a Markovian reduced dynamics may emerge.

Concretely, choosing ρ_E in (3) as an environment equilibrium state $([\rho_E\,,\,H_E] = 0)$, via the so–called *weak-coupling limit* which consists in letting $\lambda \to 0$ and $t \to +\infty$ while keeping fixed the slow time $\tau = \lambda^2\,t$, the completely positive maps Λ_t go into completely positive maps γ_t that form a semigroup: $\gamma_{t+s} = \gamma_t \circ \gamma_s$ for all $s\,,\,t \geq 0$. These maps are generated by the following time–independent master equation that is obtained from (4):

$$\partial_t \rho_S(t) = -i\,[H\,,\,\rho_S(t)] + \lambda^2\,\mathbb{D}[\rho_S(t)] =: \mathbb{L}[\rho_S(t)] \tag{6}$$

$$\mathbb{D}[\rho_S] = \sum_{i,j=1}^{d^2-1} K_{ij}\left(F_i\,\rho_S\,F_j^\dagger - \frac{1}{2}\Big(F_j^\dagger\,F_i\,\rho_S + \rho_S\,F_j^\dagger\,F_i\Big)\right). \tag{7}$$

The generator $\mathbb{L}$ consists in a Hamiltonian term with Hamiltonian operator $H = H_S + \lambda^2\,H_e$ which is the sum of the initial one in eq. (1) plus a contri-bution from the interaction of the environment. The noise and dissipative effects due to the presence of the environment are described by the linear ac-tion denoted by $\mathbb{D}$ which is written in terms of a Hilbert-Schmidt orthonor-mal basis of matrices in $M_d(\mathbb{C})$: $\mathrm{Tr}(F_j^\dagger\,F_i) = \delta_{ij}$, $F_{d^2} = 1/d$. The informa-tion relative to the presence of the environment and of its coupling (2) to the system S is contained in the $(d^2-1) \times (d^2-1)$ *Kossakowski matrix* $K = [K_{ij}]$, whose entries are sums of time–Fourier transforms of the environment 2–point time–correlation functions $\text{Tr}_E\left(\rho_E\, e^{i\,H_E\,t}\,X_\alpha^E\, e^{-i\,H_E\,t}\,X_\beta^E\right)$ evaluated at differences of the eigenvalues of the system Hamiltonian H_S.

Complete positivity, which guarantees the physical consistency [11] of the semigroup of maps γ_t generated by (6), is equivalent to the positivity of the Kossakowski matrix $K \geq 0$.

2.1. 2 and 3 open qubits

As a concrete application, we shall consider composite systems consisting either of 2 $(d = 4)$ or 3 $(d = 8)$ two–level quantum systems (qubits) weakly interacting with an environment through an interaction Hamiltonian (see (2))

$$H_I = \sum_{i=1}^{3} \Sigma_i \otimes X_E \ , \qquad \Sigma_i = \sum_{a=1}^{2,3} \sigma_i^{(a)} \ , \tag{8}$$

where $\sigma_i^{(a)}$, $i = 1,2,3$ are the Pauli matrices for the a-th qubit, $a = 1,2,3$ Namely, the form of the coupling to the environment does not distinguish between the qubits. The resulting master equation will be completely symmetric with respect to the qubits involved by further considering a context in which there are no direct interactions between the qubits and rotate with equal frequency ω around the z–direction in space:

$$H_S = \frac{\omega}{2}\left(\sigma_3^{(1)} + \sigma_3^{(2)} + \sigma_3^{(3)}\right) = \frac{\omega}{2}\Sigma_3 \ . \tag{9}$$

Thus, in the following we shall thus mainly concentrate on master equations of the highly symmetric form

$$\partial_t \rho_t = -i\frac{\omega}{2}\,[\Sigma_3\,,\,\rho_t] + \underbrace{\sum_{i,j=1}^{3} A_{ij}\left(\Sigma_i\,\rho_t\,\Sigma_j - \frac{1}{2}\{\Sigma_j\Sigma_i\,,\,\rho_t\}\right)}_{\mathbb{D}[\rho_t]} \tag{10}$$

with positive 3×3 Kossakowski matrix $A = [A_{ij}] \geq 0$.

3. Entanglement and dissipation

We shall be concerned with bipartite entanglement; under this caption, there go those correlations among two finite–level parties A, B described by the matrix algebra $M_{d_a d_b}(\mathbb{C}) = M_{d_a}(\mathbb{C}) \otimes M_{d_b}(\mathbb{C})$ acting on the Hilbert space $\mathbb{C}^{d_a} \otimes \mathbb{C}^{d_b}$ that are not explainable by classical probability theory. Entanglement or *non–separability* is the characteristic of those bipartite density matrices $\rho \in M_{d_a d_b}(\mathbb{C})$ that cannot be written as convex linear combinations of tensor products of density matrices $\rho^{(a)}$ of A, respectively $\rho^{(b)}$ of B. Vice versa, states of the form [3]

$$M_{d_a d_b}(\mathbb{C}) \ni \rho = \sum_{ij} \lambda_{ij} \rho_i^{(a)} \otimes \rho_j^{(b)} \,, \quad \lambda_{ij} > 0 \,, \quad \sum_{ij} \lambda_{ij} = 1 \qquad (11)$$

are called *separable*.

As already sketched in the introduction, we shall investigate the two following issues

- the generation of entanglement starting from an initially separable state;
- the fate of the entanglement of an initially entangled state.

We start by reviewing some aspects of bipartite entanglement and some related technical tools.

3.1. *Bipartite entanglement and its characterization*

When $d_a = d_b = 2$ or $d_a = 2, d_b = 3$ or $d_a = 3, d_b = 2$, a bipartite state ρ is entangled if and only if it does not remain positive under partial transposition; namely, if T transposes a matrix $x \in M_d(\mathbb{C})$ with respect to a given representation, $(T[x])_{ij} = x_{ji}$, then ρ is entangled if and only if

$$\mathrm{id} \otimes T[\rho] \not\geq 0 \,, \qquad (12)$$

where $\mathrm{id} \otimes \Lambda$ indicates that the operation Λ acts only on the second party of the compound system. Notice that if Λ is completely positive, $\mathrm{id} \otimes \Lambda$ is automatically positive; indeed, transposition is the prototype of positive, but not completely positive maps. The reason why in low dimension transposition is a complete witness of bipartite entanglement resides in the fact that [3] 1) all bipartite entangled states are witnessed by positive, but not completely positive maps and 2) in low dimension all positive maps are sums of completely positive maps and completely positive maps composed with transposition. On the contrary, already when $d_a = d_b = 3$, there are entangled states which remain positive under partial transposition.

Concurrence The amount of entanglement in a two-qubit state ρ can be measured by means of the concurrence.[15] Consider a two qubit pure state

$$|\psi\rangle = \alpha |00\rangle + \beta |01\rangle + \gamma |10\rangle + \delta |11\rangle$$

where $\sigma_3 |0\rangle = |0\rangle$, $\sigma_3 |1\rangle = -|1\rangle$, with $|\alpha|^2 + |\beta|^2 + |\gamma|^2 + |\delta|^2 = 1$. If it is separable, the reduced density matrix $\rho_1 = \mathrm{Tr}_2(|\psi\rangle\langle\psi|)$ obtained by partial tracing over the Hilbert space of the second qubit,

$$\rho_1 = \begin{pmatrix} |\alpha|^2 + |\beta|^2 & \alpha\gamma^* + \beta\delta^* \\ \alpha^*\gamma + \beta^*\delta & |\gamma|^2 + |\delta|^2 \end{pmatrix} ,$$

is a projection and vice versa. Therefore, the closer ρ_1 is to a projection, the less entangled is $|\psi\rangle$; this vicinity is usefully measured by the so-called *entanglement of formation*,[2] that is by the von Neumann entropy of ρ_1 (or equivalenty of ρ_2 since the reduced density matrices of a bipartite pure state have the same non null eigenvalues):

$$S(\rho_1) = -\frac{1-p}{2} \log \frac{1-p}{2} - \frac{1+p}{2} \log \frac{1+p}{2} \tag{13}$$

$$p = \sqrt{1 - C^2} , \qquad 0 \leq C = 2\,|\alpha\delta - \beta\gamma| \leq 1 . \tag{14}$$

Given the two qubit vector state

$$|\tilde{\psi}\rangle = \sigma_2 \otimes \sigma_2 |\psi^*\rangle = -\alpha^* |11\rangle + \beta^* |10\rangle + \gamma^* |10\rangle - \delta^* |00\rangle ,$$

where $|\psi^*\rangle$ is the conjugate vector of $|\psi\rangle$ with respect to the basis $|0\rangle, |1\rangle$, it is immediate to check that the 4×4 matrix

$$R = |\psi\rangle\langle\psi|\tilde{\psi}\rangle\langle\tilde{\psi}| . \tag{15}$$

has $C^2 = 4\,|\alpha\delta - \beta\gamma|^2$ as positive eigenvalue. The square root of the latter is known as the *concurrence* of the pure state $|\psi\rangle$: when it is maximal, $C = 1$, then $p = 1$ and $S(\rho_1) = \log 2$ is maximal, in which case $|\psi\rangle$ is *maximally entangled*; otherwise, when $C = 0$ is minimal, then $p = 1$, $S(\rho_1) = 0$ and $|\psi\rangle$ is separable.

In conclusion, for two qubit pure states, their entanglement is consistently measured by the concurrence C. For two qubit density matrices ρ, the entanglement of formation (13) is replaced by

$$E(\rho) = \inf \left\{ \sum_i \lambda_i S(\rho_1^i) : \rho = \sum_i \lambda_i |\psi^i\rangle\langle\psi^i| \right\} , \tag{16}$$

that is by the smallest convex combination of the entanglement of formation $S(\rho_1^i) = \mathrm{Tr}_2(|\psi^i\rangle\langle\psi^i|)$ of the pure states $|\psi^i\rangle\langle\psi^i|$ in terms of which ρ can be convexly expanded as $\sum_i \lambda_i |\psi^i\rangle\langle\psi^i|$, $\lambda_i \geq 0$, $\sum_i \lambda_i = 1$. Surprisingly, the variational quantity (16) can be expressed as in (13) with the concurrence C in (14) substituted by

$$C = \max\{0, \lambda_1 - \lambda_2 - \lambda_3 - \lambda_4\} , \tag{17}$$

where $\lambda_1 \geq \lambda_2 \geq \lambda_3 \geq \lambda_4$ are the square roots of the (positive) eigenvalues of the 4×4 matrix

$$R = \rho\,\sigma_2^{(1)}\sigma_2^{(2)}\,\rho\,\sigma_2^{(1)}\sigma_2^{(2)} , \tag{18}$$

which generalizes the matrix R in (15) and similarly has non-negative eigen-values.

Example 1. In the following, we will deal with density matrices of the form

$$\rho = \begin{pmatrix} a & 0 & 0 & 0 \\ 0 & b & c & 0 \\ 0 & c & d & 0 \\ 0 & 0 & 0 & e \end{pmatrix} , \quad a, b, d, e \geq 0 , \quad a+b+d+e = 1 , \quad bd \geq c^2 ,$$

written with respect to the standard basis $|00\rangle$, $|01\rangle$, $|10\rangle$, $|11\rangle$. In such a case, the concurrence can readily be computed as

$$C(\rho) = \max\left\{0, 2(|c| - \sqrt{ae})\right\} . \tag{19}$$

Relative entropy of entanglement Given two such density matrices, their quantum relative entropy is defined by[16]

$$S(\rho_1 || \rho_2) = \mathrm{Tr}\Big(\varrho_1(\log \varrho_1 - \log \varrho_2)\Big) . \tag{20}$$

A finite level quantum system with Hamiltonian H put in contact with a heat bath at temperature $T = 1/\beta$ (Boltzmann constant $\kappa = 1$), is expected to be driven asymptotically into the thermal equilibrium state $\varrho_T = \exp(-\beta H)/Z_\beta$, $Z_\beta = \mathrm{Tr}(\exp(-\beta H))$. If this occurs under a quantum dynamical semigroup $\varrho \mapsto \varrho_t = \gamma_t[\varrho]$, that is if $\lim_{t \to +\infty} \varrho_t = \varrho_T$, then

$$\frac{1}{\beta} S(\varrho_t || \varrho_T) = -T S(\varrho_t) + \mathrm{Tr}(\varrho_t H) + T \log Z_\beta ,$$

where $S(\varrho) = -\mathrm{Tr}\varrho \log \varrho$ is the von Neumann entropy of the state ρ. Since the second term corresponds to the system's internal energy, the first two contributions give the system's free energy corresponding to the time-evolving state ρ_t:[17]

$$F(\varrho_t) = U(\varrho_t) - T S(\varrho_t) , \quad U(\varrho_t) = \mathrm{Tr}(\varrho_t H) .$$

Finally, $F(\varrho_T) = -\log Z_\beta$ implies that the quantum relative entropy is related to the difference of free energies

$$S(\varrho_t || \varrho_T) = \beta\Big(F(\varrho_t) - F(\varrho_T)\Big) .$$

Because of the second law of thermodynamics, the above quantity should be positive and its time-derivative non-positive. The first property is guaranteed by the properties of the quantum relative entropy,[16] while the second one holds true when the irreversible time-evolution is given by a Markovian semigroup, that is when $\varrho_t = \gamma_t[\varrho]$ and $\gamma_t \circ \gamma_s = \gamma_s \circ \gamma_t = \gamma_{s+t}$ for all $s, t \geq 0$. Indeed, since $\gamma_t[\varrho_T] = \varrho_T$, one derives

$$
\begin{aligned}
S(\varrho_t \,||\, \varrho_T) &= S\Big(\gamma_t[\varrho] \,||\, \gamma_t[\varrho_T]\Big) = S\Big(\gamma_{t-s} \circ \gamma_s[\varrho] \,||\, \gamma_{t-s} \circ \gamma_s[\varrho_T]\Big) \\
&\leq S(\gamma_s[\varrho] \,||\, \gamma_s[\varrho_T]) = S(\varrho_s \,||\, \varrho_T) \quad \forall\, 0 \leq s \leq t ,
\end{aligned}
$$

where the last inequality follows from the fact that the quantum relative entropy decreases under the action of completely positive trace-preserving maps.[16]

The quantum relative entropy has also been used as a possible measure of the entanglement content of a quantum state: the so-called *relative entropy of entanglement* provides a pseudo-distance between a state and the closed convex set of separable states.[18]

When the density matrix ρ is the state of, say, a bipartite quantum system, it makes sense to introduce the *relative entropy of entanglement*,

$$
E[\varrho] = \inf_{\varrho_{sep}} S(\varrho||\varrho_{sep}) , \tag{21}
$$

as a measure of the entanglement content of ρ. Indeed, the above quantity vanishes if and only if ρ is separable and can be used to measure the distance of ρ [a] from the convex set of separable states; furthermore, it cannot increase, but it at most remains constant, under the action of local operations, described by trace-preserving completely positive maps acting independently on the two parties.[19,20]

3.2. *Dissipative generation of entanglement*

We shall first consider the case of two open qubits; in general, without the symmetry assumption on their coupling to the environment as in (10). In general, the master equation has the form

$$
\partial_t \rho_t = -i\Big[H + H_{12}\,,\,\rho_t\Big] + \mathbb{D}[\rho(t)] =: \mathbb{L}[\rho(t)] \tag{22}
$$

$$
H_{12} = \sum_{i,j=1}^{3} h_{ij}^{(12)}\,\sigma_i^{(1)}\sigma_j^{(2)} \,, \quad h^{(12)} = [h_{ij}^{(12)}] \quad \text{real matrix} \tag{23}
$$

$$
\mathbb{D}[\rho_t] = \sum_{a,b=1,2}\sum_{i,j=1,2,3} K_{ij}^{(ab)}\left(\sigma_i^{(a)}\rho_t\,\sigma_j^{(b)} - \frac{1}{2}\Big\{\sigma_j^{(b)}\sigma_i^{(a)}\,,\,\rho_t\Big\}\right) , \tag{24}
$$

---

[a]The relative entropy of entanglement is not exactly a distance since it is not symmetric.

with positive Kossakowski matrix $K = \begin{pmatrix} K^{(11)} & K^{(12)} \\ K^{(21)} & K^{(22)} \end{pmatrix}$, the matrices $K^{(aa)}$ being 3×3 positive matrix themselves and $(K^{(12)})^\dagger = K^{(21)}$.

With respect to the highly symmetric case (10), the Kossakowski matrices thus consists of blocks $K^{(ab)}$ of 3×3 matrices; the diagonal ones ($a = b$) refer to single qubits and characterize their relaxation properties, whereas the off–diagonal ones ($a \neq b$) are responsible for the source of the statistical coupling of qubit a and b. One thus expects that the generation and survival of correlations among different qubits result from the prevalence of the off–diagonal statistical coupling over diagonal decoherence.

Concerning the generation of entanglement at small times, in the case of two qubits subjected to a dissipative time-evolution of the form (22), a sufficient condition is given by the following

Proposition 1. [10] A 2-qubit separable vector state $|\psi\rangle \otimes |\phi\rangle$ becomes entangled at $t \simeq 0$ under the action of the quantum dynamical semigroup generated by (22) if the following inequality holds

$$\delta = \langle u| K^{(11)} |u\rangle \langle v|(K^{(22)})^T |v\rangle - \left| \langle v| \left(\Re e(K^{(12)}) + i\, h^{(12)} \right) |u\rangle \right|^2 < 0 , \tag{25}$$

where, $\Re e(X) = \dfrac{X + X^T}{2}$, $\Im m(X) = \dfrac{X - X^T}{2i}$ and

$$u_i = \langle \psi|\sigma_i|\psi_\perp\rangle , \quad v_i = \langle \phi_\perp^*|\sigma_i|\phi^*\rangle . \tag{26}$$

The idea of the proof is as follows: First of all, one can restrict oneself to considering initial separable states that are pure. Indeed, if these kind of states cannot be entangled at small times, neither can their linear convex combinations. Then, one considers the first order expansion in t of the time-evolution generated by (22)

$$\rho_t \simeq |\psi\rangle\langle\psi| \otimes |\phi\rangle\langle\phi| + t\, \mathbb{L}[|\psi\rangle\langle\psi| \otimes |\phi\rangle\langle\phi|] .$$

Since the open quantum system of interest consists of two qubits, the possible generation of entanglement is fully witnessed by partial transposition that yields

$$\mathrm{id} \otimes T[\rho_t] \simeq |\psi\rangle\langle\psi| \otimes |\phi^*\rangle\langle\phi^*| + t\, \mathrm{id} \otimes T[\mathbb{L}[|\psi\rangle\langle\psi| \otimes |\phi\rangle\langle\phi|] .$$

The resulting state at small times is thus entangled if there are negative mean values

$$\langle\Psi|\mathrm{id} \otimes T[\rho_t]|\Psi\rangle \simeq t\, \langle\Psi|\mathrm{id} \otimes T[\mathbb{L}[|\psi\rangle\langle\psi| \otimes |\phi\rangle\langle\phi|]]|\Psi\rangle < 0$$

with respect to states $|\Psi\rangle$ orthogonal to $|\psi\rangle \otimes |\phi^*\rangle$:

$$|\Psi\rangle = \alpha|\psi\rangle \otimes |\phi_\perp^*\rangle + \beta|\psi_\perp\rangle \otimes |\phi^*\rangle + \eta|\psi_\perp\rangle \otimes |\phi_\perp^*\rangle \ ,$$

where $\langle\psi_\perp|\psi\rangle = \langle\phi_\perp^*|\phi^*\rangle = 0$ and $|\phi^*\rangle$ is the vector conjugate to $|\phi\rangle$ in the chosen representation with respect to which transposition is defined.

The choice of $|\Psi\rangle$ eliminates the zeroth–order term and must be entangled otherwise the partial transposition could be shifted to it by duality,

$$\mathrm{Tr}\Big(|\Psi\rangle\langle\Psi|\,\mathrm{id} \otimes T[\rho_t]\Big) = \mathrm{Tr}\Big(\mathrm{id} \otimes T[|\Psi\rangle\langle\Psi|]\,\rho_t\Big) \ ,$$

and the mean value would always be non–negative.

It turns out that this condition is nearly necessary, failing to be so only when the quantity δ in (25) vanishes. Indeed, one has

Proposition 2 [21]   If δ in (25) is strictly positive (> 0); then, separable two-qubit states cannot get entangled by the quantum dynamical semigroups generated by (22).

Example 3. [11]   Let us consider two identical open qubits, evolving in time according to the master equation (10) with $\Sigma_i = \sigma_i^{(1)} + \sigma_i^{(2)}$ and [b]

$$A = \begin{pmatrix} 1 & i\beta & 0 \\ -i\beta & 1 & 0 \\ 0 & 0 & 1 \end{pmatrix} \ , \quad 0 < \beta < 1 \ .$$

In such a case, the Kossakowski matrix in (24) reads $K = \begin{pmatrix} A & A \\ A & A \end{pmatrix}$; thus, or the initial state $|0\rangle \otimes |1\rangle$, where $\sigma_3|0\rangle = |0\rangle$, $\sigma_3|1\rangle = -|1\rangle$, one computes $|u\rangle = |v\rangle = (\langle 0|\sigma_i|1\rangle) = (1, -i, 0)$ so that (25) reads

$$\delta = \langle u|A|u\rangle\,\langle u|A^T|u\rangle - \left|\langle u|\frac{A+A^T}{2}|u\rangle\right|^2 = -\left|\langle u|\frac{A-A^T}{2}|u\rangle\right|^2$$
$$= -\beta^2 < 0 \ .$$

In the following example, rather than with the generation of entanglement at small times, we deal with its asymptotic behavior.

Example 4. [22]   The master equation (10) with Kossakovski matrix as in the previous example can be explicitly solved by representing the time–evolving state $\rho_t = \sum_{i,j=1}^4 \rho_{ij}(t)\,|i\rangle\langle j|$ with respect to the orthonormal basis:

$$|1\rangle = |00\rangle \ , \ |2\rangle = |11\rangle \ |3\rangle = \frac{|01\rangle + |10\rangle}{\sqrt{2}} \ , \ |3\rangle = \frac{|01\rangle - |10\rangle}{\sqrt{2}} \ ,$$

---

[b]The microscopic justification of such a form will become apparent in Section 5.

where $|ij\rangle$, $i,j = 0,1$, and $\sigma_3|j\rangle = (-1)^j|j\rangle$.[22]

Any initial state diagonal with respect to this basis,

$$\rho = a\,|1\rangle\langle 1| + d\,|2\rangle\langle 2| + b\,|3\rangle\langle 3| + c\,|4\rangle\langle 4| = \begin{pmatrix} a & 0 & 0 & 0 \\ 0 & \frac{b+c}{2} & \frac{b-c}{2} & 0 \\ 0 & \frac{b-c}{2} & \frac{b+c}{2} & 0 \\ 0 & 0 & 0 & d \end{pmatrix}, \qquad (27)$$

remains diagonal under the dissipative time–evolution generated by (10) and the resulting asymptotic state ρ_∞ can explicitly be computed:

$$\rho \mapsto \rho_t = a_t\,|1\rangle\langle 1| + d_t\,|2\rangle\langle 2| + b_t\,|3\rangle\langle 3| + c\,|4\rangle\langle 4| \qquad (28)$$

$$\rho_\infty = \frac{\beta^2}{3+\beta^2}\,(1-c)\,|1\rangle\langle 1| + \frac{(1+\beta)^2}{3+\beta^2}\,(1-c)\,|2\rangle\langle 2| \qquad (29)$$

$$+ \frac{(1-\beta^2)}{3+\beta^2}\,(1-c)\,|3\rangle\langle 3| + c\,|4\rangle\langle 4|\ .$$

The concurrence can thus be computed by using (19):

$$C(\rho_t) = \max\left\{0, 2\left(\frac{|b_t - c|}{2} - \sqrt{a_t\,d_t}\right)\right\} \qquad (30)$$

$$C(\rho_\infty) = \max\left\{0, \frac{\left|1 - \beta^2 - 4c\right| - 2(1-\beta^2)(1-c)}{3+\beta^2}\right\}. \qquad (31)$$

4. Entropy and entanglement production

Based on the thermodynamical arguments of Section 3, one may consider generic open quantum dynamics $\varrho \mapsto \gamma_t[\varrho] = \varrho_t$ with asymptotic states $\lim_{t\to+\infty}\varrho_t = \varrho_\infty$, that are not necessarily thermal ones. The speed of convergence to such stationary states starting from an initial state ϱ can then be measured by the *entropy rate*

$$\sigma[\varrho_t] = -\frac{d}{dt}S(\varrho_t||\varrho_\infty) = \mathrm{Tr}\Big(\dot\varrho_t(\log\varrho_\infty - \log\varrho_t)\Big). \qquad (32)$$

Example 5. The entropy production that accompanies the tendency to equilibrium of the states of Example 4, can explicitly be computed; indeed, being the states ϱ_t and ϱ_∞ diagonal with respect to the same orthonormal basis, the entropy rate (32) has the analytic expression

$$\sigma[\varrho_t] = \dot a_t\,\log\frac{(1-\beta)^2(1-c)}{a_t(3+\beta^2)} + \dot b_t\,\log\frac{(1-\beta^2)(1-c)}{b_t(3+\beta^2)}$$

$$+ \dot d_t\,\log\frac{(1+\beta)^2(1-c)}{d_t(3+\beta^2)}\ . \qquad (33)$$

Analogously, one may look at the *entanglement rate* when the system evolves, i.e. at the time-derivative of the pseudo-distance given by the relative entropy of entanglement (21):

$$\sigma_E[\varrho_t] = \frac{d}{dt} E[\varrho_t] \ . \tag{34}$$

Example 6. [23]   In the case of the diagonal states (27) of Example 4, the variational computation involved by the definition (21) can be solved and the states achieving the supremum computed analytically. It is achieved at the separable states

$$\rho_{sep} = x \, |1\rangle\langle 1| + u \, |3\rangle\langle 3| + v \, |4\rangle\langle 4| + y \, |2\rangle\langle 2|$$

$$x, y, u, v \in \mathbb{R} \ , \quad x + u + v + y = 1 \ , \quad \frac{|u - v|}{2} \le \sqrt{xy} \ ,$$

and it is given by explicitly computing [23] the supremum in

$$E(\rho_t) = -S(\rho_t) - \sup_{x,y,u,v} \Big(a_t \log x \, + \, d_t \, \log y \, + \, b_t \log u \, + c \log v \Big) \ ,$$

where a_t, b_t, c_t are the coefficients appearing in (28).

In Ref. 14 it is argued that

$$\Big| \sigma_E[\varrho_t] \Big| \le \sigma[\varrho_t] \tag{35}$$

always holds in absence of direct entangling interactions between the parties. The argument on which this conjecture is based is that decoherence is expected to deplete entanglement before reaching the asymptotic state and thus before the entropy production vanishes.

By using (28), (29) and (19) to compute the concurrences and the explicit analytical results of Example 5 and 6 to plot (32) and (34), such a conjecture can be put to test. It turns out to be false in cases when the asymptotic states are entangled.

Case 1.   Initial and final states: separable. The initial state, in the diagonal form (27),

$$\rho = \frac{1}{2} |2\rangle\langle 2| \, + \, \frac{1}{10} \, |3\rangle\langle 3| \, + \, \frac{2}{5} \, |4\rangle\langle 4|$$

is mixed and entangled and goes into the separable asymptotic state

$$\rho_\infty = \frac{3(1-\beta)^2}{5(3+\beta^2)} \, |1\rangle\langle 1| + \frac{3(1+\beta)^2}{5(3+\beta^2)} \, |2\rangle\langle 2| + \frac{3(1-\beta^2)}{5(3+\beta^2)} \, |3\rangle\langle 3| \, + \, \frac{2}{5} \, |4\rangle\langle 4| \ .$$

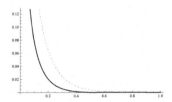

Figure 1.   Case 1: $\beta = 0.5$; $\sigma[\rho_t]$ dashed line, $|\sigma_E[\rho_t]|$ continuous line

For $\beta = 0.5$, the initial and final concurrences are $C(\rho) = \frac{3}{5}$ and $C(\rho_\infty) = 0$. The conjecture (35) is confirmed in such a case.

Case 2.   Initial and final states: entangled. The initial state

$$\rho = \frac{3}{10}|2\rangle\langle 2| + \frac{1}{10}|3\rangle\langle 3| + \frac{3}{5}|4\rangle\langle 4|$$

is mixed and entangled and goes into the entangled asymptotic state

$$\rho_\infty = \frac{2(1-\beta)^2}{5(3+\beta^2)}|1\rangle\langle 1| + \frac{2(1+\beta^2)}{5(3+\beta^2)}|2\rangle\langle 2| + \frac{2(1-\beta^2)}{5(3+\beta^2)}|3\rangle\langle 3| + \frac{3}{5}|4\rangle\langle 4| \ .$$

The concurrences are $C(\rho) = \frac{1}{2}$ and $C(\rho_\infty) = \frac{3(1+3\beta^2)}{5(3+\beta^2)}$: with $\beta = 0.5$ the final entanglement is smaller than the initial one, $C(\rho_\infty) < C(\rho) = \frac{1}{2}$. The following graph shows that, in such a case, the conjecture (35) is always violated.

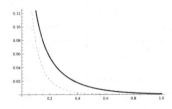

Figure 2.   Case 2: $\beta = 0.5$; $\sigma[\rho_t]$ dashed line, $|\sigma_E[\rho_t]|$ continuous line

Case 3.   Final entanglement larger than the initial one. Initial state as in Case 2, but $\beta = .8$: $C(\rho_\infty) > C(\rho) = \frac{1}{2}$. In this case, the following graph shows that the conjecture (35) is violated after some time.[c]

---

[c]The cusp is due to the absolute value in definition (34).

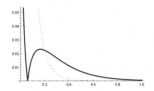

Figure 3.   Case 3: $\beta = 0.8$; $\sigma[\varrho_t]$ dashed line, $|\sigma_E[\varrho_t]|$ continuous line

5. Three open qubits in a symmetric environment

In the second part of this contribution, we consider 3 identical qubits in a symmetric environment with dissipative time-evolution generated by a master equation of the form (10). Unlike in Example 3, we shall consider the following Kossakowski matrix

$$A = \begin{pmatrix} a & ib & 0 \\ -ib & a & 0 \\ 0 & 0 & c \end{pmatrix} , \quad c \geq 0 , \ a^2 \geq b^2 . \tag{36}$$

It corresponds to an interaction Hamiltonian [24]

$$H_I = \sum_{i=1}^{3} \sum_{a=1}^{2,3} \sigma_i^{(a)} \otimes X_{a,i}^E ,$$

with an environment consisting of scalar Bosons with thermal two-point correlation functions

$$\mathrm{Tr}_E \left(X_{a,i}^E \, X_{b,j}^E(t) \right) = \delta_{ij} \int \mathrm{d}\vec{x} \, \mathrm{d}\vec{y} \, f^{(a)}(\vec{x}) \, f^{(b)}(\vec{y}) \, G_\beta(t, \vec{x} - \vec{y})$$

$$G_\beta(t, \vec{x}) = \int \frac{\mathrm{d}^4 k}{2\pi} \, \theta(k_0) \, \delta(k^2) \left(\frac{e^{-i t \, k_0 + i \vec{k} \cdot \vec{x}}}{1 - e^{-\beta \, k_0}} + \frac{e^{i t \, k_0 - i \vec{k} \cdot \vec{x}}}{e^{\beta \, k_0} - 1} \right) ,$$

in the limit where the smearing functions $f^{(a,b)}(\vec{x})$ spatially localize the two qubits at the origin. The coefficients of the Kossakowski matrix are then given by

$$a = \frac{\omega}{4\pi} \coth(\beta \omega / 2) , \quad b = -\frac{\omega}{4\pi} , \quad c = \frac{1}{2\pi\beta}$$

The case of Example 3 corresponds to the small β (high temperature) approximation of the present case.

As announced in the Introduction, purpose of this section is to apply the following protocol to two equal qubits symmetrically interacting with an environment as described by (10) with Kossakowski matrix as in (36) :[32]

- Add a third qubit equal to the given two qubits;
- switch on the symmetric coupling to the environment;
- let equilibrium be reached;
- eliminate one qubit.

Then, we show that the remaining two qubits can

(1) get entangled when they would not if directly immersed in the symmetric environment;
(2) show an asymptotic entanglement gain with respect to the initial state, while by direct immersion they would not;
(3) show a greater entanglement gain with respect to what they would exhibit by direct immersion.

Thus, we shall essentially be concerned with the stationary states of (10) for two ($\Sigma_i = \sigma_i^{(1)} + \sigma_i^{(2)}$) and three qubits ($\Sigma_i = \sigma_i^{(1)} + \sigma_i^{(2)} + \sigma_i^{(3)}$).

Example 7. One qubit: $\Sigma_i = \sigma_i$. The stationary states of

$$\partial_t \rho_t = -i\frac{\omega}{2}\left[\sigma_3\,,\,\rho_t\right] + \sum_{i,j=1}^{3} K_{ij}\left(\sigma_i\,\rho_t\,\sigma_j - \frac{1}{2}\left\{\sigma_j\,\sigma_i\,,\,\rho_t\right\}\right)\,,$$

can be evaluated by translating it into an equation

$$\dot{\vec{r}}_t = \mathcal{L}\,\vec{r}_t - \vec{z}\,, \qquad \mathcal{L} = \begin{pmatrix} a+c & -\omega/2 & 0 \\ \omega/2 & a+c & 0 \\ 0 & 0 & 2a \end{pmatrix}\,, \qquad \vec{z} = \begin{pmatrix} 0 \\ 0 \\ 2b \end{pmatrix}$$

for the Bloch vector $\vec{r}_t$ of the density matrix $\rho_t = \frac{1}{2}(1 + \vec{r}_t \cdot \vec{\sigma})$. It turns out that there is only one (full rank) stationary state:

$$\rho_\infty^* = \frac{1}{2}\left(1 + r_\infty\,\sigma_3\right)\,, \qquad \vec{r}_\infty = \begin{pmatrix} 0 \\ 0 \\ r_\infty \end{pmatrix} = \mathcal{L}^{-1}\vec{z} = \begin{pmatrix} 0 \\ 0 \\ b/a \end{pmatrix}\,.$$

5.1. *Stationary states: general results*

An abstract approach to the characterization of the manifold of stationary states of quantum dynamical semigroups and of the tendency to equilibrium were developed long ago at the beginning of the theory of open quantum systems.[4,17,25–27] Only a partial characterization was then achieved; recently, the issue has been taken on again because of its increasing importance in quantum information theory.[28–31]

We shall hereby review some tools that are necessary for the rest of this contribution. In order to inspect the manifold of invariant states of a given quantum dynamical semigroup, it is convenient to consider the dual (Heisenberg) time-evolution, $M_d(\mathbb{C}) \ni x \mapsto \gamma_t^T[x] \in M_d(\mathbb{C})$, of the system observables (matrices $x \in M_d(\mathbb{C})$) and to write the corresponding master equation in Kraus diagonal form :[4]

$$\partial_t x_t = i\Big[H \, , \, x_t\Big] + \sum_{i=1}^{3} \Big(V_i^\dagger \, x_t \, V_i - \frac{1}{2}\big\{V_i^\dagger \, V_i \, , \, x_t\big\}\Big) = \mathbb{L}^T[x] \, , \qquad (37)$$

whose generator $\mathbb{L}^T$ (dual of $\mathbb{L}$ in (22)) is obtained by differentiating the duality relation

$$\mathrm{Tr}\Big(x \, \gamma_t[\rho]\Big) = \mathrm{Tr}\Big(\gamma_t^T[x] \, \rho\Big) \, .$$

Example 8.   In the case of the master equation (10) for identical qubits, the generator in the Heisenberg picture reads

$$\partial_t x_t = i\frac{\omega}{2}\Big[\Sigma_3 \, , \, x_t\Big] + \sum_{i,j=1}^{3} A_{ij}\Big(\Sigma_j x_t \Sigma_i - \frac{1}{2}\{\Sigma_j \Sigma_i \, , \, x_t\}\Big) \, .$$

By diagonalizing the Kossakowski matrix (36), one gets the diagonal form

$$\partial_t x_t = i\frac{\omega}{2}\Big[\Sigma_3 \, , \, x_t\Big] + \sum_{i=1}^{3} \Big(V_i^\dagger \, x_t \, V_i - \frac{1}{2}\big\{V_i^\dagger \, V_i \, , \, x_t\big\}\Big) \, , \qquad (38)$$

with Kraus operators

$$V_{1,2}^\dagger = \sqrt{2(a \mp b)}\frac{\Sigma_1 \mp i\Sigma_2}{2} \, , \qquad V_3 = \sqrt{c}\,\Sigma_3 \, . \qquad (39)$$

The following one is an important structural result given to [27] that we rephrase for finite level systems.

Theorem 1.   Consider the set of γ_t^T-invariant $d \times d$ matrices

$$M_d(\mathbb{C}) \supseteq M_\gamma := \Big\{x \in M_d(\mathbb{C}) \, : \, \gamma_t^T[x] = x\Big\} \, ;$$

if ρ_0 is a full rank stationary state ($\gamma_t[\rho_0] = \rho_0$), then M_γ is a $*$ sub–algebra. Namely, $x \in M_\gamma \Rightarrow x^\dagger \in M_\gamma$, $x^\dagger x \in M_\gamma$ and

$$x \in M_\gamma \Rightarrow \gamma_t^T[x] = x \, , \; \gamma_t^T[x^\dagger] = x^\dagger \, , \; \gamma_t^T[x^\dagger x] = x^\dagger x \, .$$

Furthermore, the time-average

$$\mathbb{E}[x] = \lim_{T \to +\infty} \frac{1}{T}\int_0^T \mathrm{d}t \, \gamma_t^T[x] \qquad (40)$$

defines a unique *conditional expectation*: $\mathbb{E} : M_d(\mathbb{C}) \mapsto M_\gamma$ such that

$$\mathbb{E}[y\,x\,z] = y\,\mathbb{E}[x]\,z \qquad \forall x \in M_d(\mathbb{C}) \ , \ \forall y, z \in M_\gamma \ . \tag{41}$$

Notice that, given $\mathbb{E}$, by duality, one gets a completely positive map $\mathbb{F}$ onto the γ_t–invariant states; this map associates to any state ρ an invariant state $\rho_\infty = \mathbb{F}[\rho]$; as follows

$$\mathrm{Tr}\Big(\rho\,\mathbb{E}[x]\Big) = \mathrm{Tr}\Big(\mathbb{F}[\rho]\,x\Big) \ , \qquad \forall x \in M_d(\mathbb{C}) \ . \tag{42}$$

If a full rank γ_t–stationary state exists, to construct the γ_t^{T}–invariant sub–algebra M_γ one looks at the so–called *commutant set* $\{V_i\}'$ of the Kraus set $\{V_i\}$:

$$\{V_i\}' := \Big\{x \in M_d(\mathbb{C}) \ : \ [x, V_i] = 0 \,\forall i\Big\} \ . \tag{43}$$

Let us consider the so–called *dissipative set*

$$D_\gamma := \Big\{x \in M_d(\mathbb{C}) \ : \ \mathbb{L}^{\mathrm{T}}[x^\dagger x] - x^\dagger \mathbb{L}^{\mathrm{T}}[x] - \mathbb{L}^{\mathrm{T}}[x^\dagger]x = 0\Big\} \ .$$

This set equals the commutant set: $\{V_i\}' \subseteq D_\gamma$ is obvious, while

$$x \in D_\gamma \Rightarrow \sum_i [V_i \, , \, x]^\dagger\,[V_i \, , \, x] = 0 \Rightarrow [x \, , \, V_i] = 0 \ .$$

Observe that $M_\gamma \subseteq D_\gamma$: this follows from the definition of the dissipative set and from the fact that

$$x \in M_\gamma \Rightarrow \gamma_t^{\mathrm{T}}[x] = x \ , \ \gamma_t^{\mathrm{T}}[x^\dagger] = x^\dagger \ , \ \gamma_t^{\mathrm{T}}[x^\dagger x] = x^\dagger x$$

so that, by differentiation,

$$x \in M_\gamma \Rightarrow \mathbb{L}^{\mathrm{T}}[x] = 0 \ , \ \mathbb{L}^{\mathrm{T}}[x^\dagger] = 0 \ , \ \mathbb{L}^{\mathrm{T}}[x^\dagger\,x] = 0 \ .$$

Consider now the commutant set (it is actually a sub–algebra in such a case) $\{H, V_i, V_i^\dagger\}'$: this is contained in M_γ for $\{H, V_i, V_i^\dagger\} \supset \{V_i\}$.

Lemma 1. If $\{V_i\}' = \{H, V_i, V_i^\dagger\}'$ then $M_\gamma = \{V_i\}'$.

Proof: $M_\gamma \subseteq D_\gamma = \{V_i\}' = \{H, V_i, V_i^\dagger\}' \subseteq M_\gamma$.

Example 9. A quantum dynamical semigroup γ_t has a unique stationary state if $\{V_i\}' = \{H, V_i, V_i^\dagger\}' = \{\lambda\mathbb{1}\} = M_\gamma$. In fact, in such a case,

$$M_\gamma = \{\lambda\mathbb{1}\} \Longrightarrow \mathbb{E}[x] = \lambda(x)\,\mathbb{1} \ .$$

Notice that, while the map $\mathbb{F}$ in (42) associates to any state ρ a stationary state $\rho_\infty = \mathbb{F}[\rho]$, it is not in general true that $\gamma_t[\rho] \to \rho_\infty$ when

$t \to +\infty$. In [27] it was proved that such is the case when the invariant sub–algebra M_γ coincides with the subset

$$N_\gamma := \left\{ x \in M \; : \; \gamma_t^{\mathrm{T}}[x^\dagger \, x] = \gamma_t^{\mathrm{T}}[x^\dagger] \, \gamma_t^{\mathrm{T}}[x] \; \forall \, t \geq 0 \right\} .$$

By differentiation at $t = 0$, this subset is seen to be contained in D_γ. Also, it contains the sub–algebra $\{H, V_i, V_i^\dagger\}'$. Thus,

Lemma 2.   If $\{V_i\}' = \{H, V_i, V_i^\dagger\}'$, $\lim_{t \to +\infty} \gamma_t[\rho] = \mathbb{F}[\rho]$ for all $\rho \in M_d(\mathbb{C})$.

Proof: From Lemma 1 and the previous considerations it follows that $M_\gamma = \{V_i\}' = \{H, V_i, V_i^\dagger\}' \subseteq N_\gamma \subseteq D_\gamma = \{V_i\}'$.

In the following discussion, the context will be such that Lemmas 1 and 2 can be applied.

Application to two identical qubits In this section we apply the abstract theory outlined above to the concrete case of two open identical qubits in a symmetric environment whose time-evolution is given by (10) with Kossakowski matrix as in (36).

One can directly check that a full rank stationary two qubit state is given by $\rho_\infty^{\otimes 2} = \rho_\infty^* \otimes \rho_\infty^*$ where ρ^* is the one–qubit stationary state of Example 7. Then, notice that the commutant set $\{V_i\}' = \{\Sigma_i\}' = \{H, V_i, V_i^\dagger\}'$ so that the γ_t^{T}–invariant sub–algebra M_γ equals the commutant set. In order to construct it, one writes

$$M_4(\mathbb{C}) \ni x = \lambda \mathbb{1} + \sum_{i=1}^{3} \sum_{a=1}^{2} \lambda_i^{(a)} \, \sigma_i^{(a)} + \sum_{i,j=1}^{3} \lambda_{ij} \, \sigma_i \otimes \sigma_j$$

and imposes that $[x \, , \, \Sigma_p] = 0$ for all $p = 1, 2, 3$, whence

$$M_\gamma = D_\gamma = \{\Sigma_i\}' = \mathrm{LinSpan}\left\{ \mathbb{1} \, , \, T = \sum_{j=1}^{3} \sigma_j \otimes \sigma_j \right\} .$$

Therefore, M_γ is a commutative algebra generated by the minimal orthogonal projections

$$P = \frac{1}{4}\left(\mathbb{1} - T \right) , \quad Q = \mathbb{1} - P = \frac{1}{4}\left(3 + T \right) , \tag{44}$$

where P projects onto the singlet-state

$$P = |\Psi\rangle\langle\Psi| \, , \, |\Psi\rangle = \frac{1}{\sqrt{2}}\left(|0\rangle \otimes |1\rangle - |1\rangle \otimes |0\rangle \right) .$$

Then, the conditional expectation onto the sub–algebra M_γ is easily constructed by setting $\mathbb{E}[x] = \lambda(x)\,P + \mu(x)\,Q$; indeed, from (40) one derives

$$\mathbb{E}[P\,x\,P] = P\,\mathbb{E}[x]\,P = \lambda(x)\,P \; , \quad \mathbb{E}[Q\,x\,Q] = Q\,\mathbb{E}[x]\,Q = \mu(x)\,Q \; , \quad (45)$$

whence

$$\lambda(x) = \frac{\mathrm{Tr}\Big(P\,\rho_\infty^{\otimes 2}\,P\,x\Big)}{\mathrm{Tr}\Big(\rho_\infty^{\otimes 2}\,P\Big)} \; , \quad \mu(x) = \frac{\mathrm{Tr}\Big(Q\,\rho_\infty^{\otimes 2}\,Q\,x\Big)}{\mathrm{Tr}\Big(\rho_\infty^{\otimes 2}\,Q\Big)} \; . \quad (46)$$

Finally, by the duality relation (42), one finds that the stationary states are obtained as

$$\rho_\infty = \mathbb{F}[\rho] = \frac{4\Big(1 - \mathrm{Tr}\Big(\rho\,P\Big)\Big)}{3 + r_\infty^2}\,\rho_\infty^{\otimes 2} + \frac{4\mathrm{Tr}\Big(\rho\,P\Big) - 1 + r_\infty^2}{3 + r_\infty^2}\,P \; . \quad (47)$$

Lemma 2 holds in this context; therefore, any initial ρ will tend to ρ_∞ as $t \to +\infty$.

Example 10. Consider the following class of two–qubit states:

$$\rho(\alpha) = \alpha\,\mathbb{1} + (1 - 4\,\alpha)\,P \; , \quad 0 \le \alpha \le 1/3 \; ; \quad (48)$$

applying (47) one gets the stationary states

$$\rho_\infty(\alpha) = \frac{12\alpha}{3 + r_\infty^2}\,\rho_\infty^{\otimes 2} + \frac{3 + r_\infty^2 - 12\alpha}{3 + r_\infty^2}\,P \; , \quad (49)$$

whose concurrence (17) can be explicitly computed:

$$C(\rho_\infty(\alpha)) = \frac{1}{2} - 3\alpha\,\frac{3 - r_\infty^2}{3 + r_\infty^2} > 0$$

for $0 \le \alpha < \alpha(r_\infty) = \dfrac{3 + r_\infty^2}{6(3 - r_\infty^2)}$.

For the states $\rho(\alpha)$ and the associated stationary states $\rho_\infty(\alpha)$, the following possibilities arise:

(1) $\rho(\alpha)$ and $\rho_\infty(\alpha)$ are both separable if $\dfrac{1}{6} \le \alpha(r_\infty) \le \alpha \le \dfrac{1}{3}$;

(2) $\rho(\alpha)$ is separable and $\rho_\infty(\alpha)$ entangled if $\dfrac{1}{6} \le \alpha \le \alpha(r_\infty)$;

(3) $\rho(\alpha)$ and $\rho_\infty(\alpha)$ are both entangled if $0 \le \alpha < 1/6$;

(4) Define

$$\Delta(\alpha) := C(\rho_\infty(\alpha)) - C(\rho(\alpha)) = 9\,\alpha\,\frac{1 + r_\infty^2}{3 + r_\infty^2} - \frac{1}{2} \; ; \quad (50)$$

then there is an *entanglement gain*, namely $\Delta(\alpha) > 0$, if

$$\alpha > \alpha^*(r_\infty) = \frac{3 + r_\infty^2}{18(1 + r_\infty^2)} \; .$$

Application to three identical qubits We now take on the case of three identical qubits in the symmetric environment described by (10) and (36). As $\rho_\infty^{\otimes 2} = \rho_\infty^* \otimes \rho_\infty^*$ where ρ^* is the one–qubit stationary state of Example 7 is a full rank stationary state for two qubits, such is $\rho_\infty^{\otimes 3} = \rho_\infty^* \otimes \rho_\infty^* \otimes \rho_\infty^*$ for three qubits. Furthermore, it also holds that $\{V_i\}' = \{\Sigma_i\}' = \{H, V_i, V_i^\dagger\}'$ where $\Sigma_i = \sigma_i^{(1)} + \sigma_i^{(2)} + \sigma_i^{(3)}$; thus $M_\gamma = \{V_i\}'$.

In order to inspect the structure of M_γ in such a case, write

$$M_8(\mathbb{C}) \ni x = \lambda \mathbb{1} + \sum_{i=1}^{3}\sum_{a=1}^{3} \lambda_i^{(a)} \sigma_i^{(a)} + \sum_{i,j=1}^{3}\sum_{a \neq b} \lambda_{ij}^{(ab)} \sigma_i^{(a)} \otimes \sigma_j^{(b)} \quad (51)$$

$$+ \sum_{i,j,k=1}^{3} \lambda_{ijk}\sigma_i^{(1)}\sigma_j^{(2)}\sigma_k^{(3)} \quad (52)$$

and imposes $[x\,,\,\Sigma_p] = 0$ for all $p = 1, 2, 3$. This yields

$$M_\gamma = \{\Sigma_i\}' = \mathrm{LinSpan}\Big\{\mathbb{1}\,,\,S^{(ab)}\,,\,S\Big\}\,,\,S^{(ab)} = \sum_{i=1}^{3}\sigma_i^{(a)}\sigma_i^{(b)}$$

$$S = \sum_{ijk} \epsilon_{ijk}\sigma_i^{(1)}\sigma_j^{(2)}\sigma_k^{3)} \; .$$

The γ_t^T–invariant sub–algebra is thus non–commutative and the explicit construction of the conditional expectation $\mathbb{E}$ (40) onto it rather difficult. We content ourselves with the following two actions of the dual map $\mathbb{F}$ (42):

$$\mathbb{F}[P \otimes \mathbb{1}] = 2\,P \otimes \rho_\infty^* \quad (53)$$

$$\mathbb{F}[\mathbb{1}] = \frac{8}{1 + r_\infty^2} \rho_\infty^{\otimes 3} + \frac{8r_\infty^2}{3(1 + r_\infty^2)} \sum_{a<b=2}^{3} P^{(ab)}(\rho_\infty^*)_{(c)}\,, \quad (54)$$

where P is the projector (44) onto the singlet state of the first two qubits, $P^{(ab)}$ are the same for qubits a and b, while $(\rho_\infty^*)_{(c)}$ denotes the state ρ_∞^* of qubit c.

These expressions suffice to find all stationary states associated to an extension to three qubits of the special class of two qubit–states (48).

Example 11. Add a totally depolarized third qubit to the two–qubit states $\rho(\alpha)$ in (48): $\rho(\alpha) \to \rho_{123}(\alpha)$, where

$$\rho_{123}(\alpha) = \rho(\alpha) \otimes \frac{\mathbb{1}}{2} = \frac{\alpha}{2}\mathbb{1} + \frac{1 - 4\,\alpha}{2} P \otimes \mathbb{1} \; . \quad (55)$$

Then, use (53) and (54) to associate to each of them the stationary state

$$\rho_\infty^{123}(\alpha) = \mathbb{F}[\rho^{123}(\alpha)] = \frac{4\,\alpha}{1+r_\infty^2}\,\rho_\infty^{\otimes 3} + \frac{4\,\alpha\,r_\infty^2}{3(1+r_\infty^2)} \sum_{a<b=2}^3 P^{(ab)}(\rho_\infty^*)_{(c)}$$
$$+ (1-4\,\alpha)\,P^{(12)}(\rho_\infty^*)_3 \ . \tag{56}$$

As for two qubits, Lemma 2 holds in the present context, therefore $\displaystyle\lim_{t\to+\infty}\gamma_t[\rho_{123}(\alpha)] = \rho_\infty^{123}(\alpha)$.

5.2. *Gaining entanglement by dissipation*

We can enforce the protocol outlined in the previous Section and show its properties by using the states (48).

We start by adding a totally depolarized third qubit to $\rho(\alpha)$, thus constructing the three–qubit states $\rho_{123}(\alpha)$ in (55).

Then, the 3 qubits prepared in such a state are symmetrically coupled to an environment giving rise to the dissipative time-evolution generated by (10) with Kossakowski matrix (36).

After the three qubits have reached the stationary state $\rho_\infty^{123}(\alpha)$ in (56), the third qubit is eliminated thus leading to the two–qubit state

$$\rho_\infty^{12}(\alpha) = \mathrm{Tr}_3\left(\rho_\infty^{123}(\alpha)\right)$$
$$= \frac{4\,\alpha}{1+r_\infty^2}\,\rho_\infty^* \otimes \rho_\infty^* + \frac{4\,\alpha\,r_\infty^2 + 3(1-4\,\alpha)(1+r_\infty^2)}{3(1+r_\infty^2)}\,P$$
$$+ \frac{2\,\alpha\,r_\infty^2}{3(1+r_\infty^2)}\left(\mathbb{1} \otimes \rho_\infty^* + \rho_\infty^* \otimes \mathbb{1}\right) \ . \tag{57}$$

One can thus compare the asymptotic entanglement content of the two–qubit states $\rho_{12}(\alpha)$ with respect to the states $\rho_\infty(\alpha)$ in (49) when they are not subjected to the protocol. We shall consider:

(1) the concurrence $C(\rho_\infty^{12}(\alpha))$ when $C(\rho_\infty(\alpha)) = 0$;
(2) the entanglement gain (50) after application of the protocol,

$$\Delta_{12}(\alpha) := C(\rho_\infty^{12}(\alpha)) - C(\rho(\alpha)) \ ,$$

when, without protocol, $\Delta(\alpha) = C(\rho_\infty(\alpha)) - C(\rho(\alpha)) \le 0$;
(3) the entanglement gain (50) after applying the protocol when, without protocol, $\Delta(\alpha) = C(\rho_\infty(\alpha)) - C(\rho(\alpha)) > 0$.

We show below three plots [32] where the independent variables are the parameters α, relative to the states in the chosen class and r_∞ determined

by the environment. Concerning the first point above, the very much local-
ized peak in the firs graph shows that for a small subset of two–qubit states
$\rho(\alpha)$ and of symmetric environemnts, the protocol can indeed provide more
entanglement with respect to direct immersion.

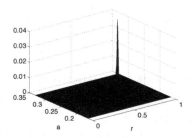

Figure 4. $C(\rho_\infty^{12}(\alpha))$ when $C(\rho_\infty(\alpha)) = 0$

The benefits coming from the protocol are more clearly visible by considering the entanglement gain: the following two graphs exhibits the areas
where the difference $\Delta_{12}(\alpha) - \Delta(\alpha)$ is positive in relations to points (2) and
(3) above.

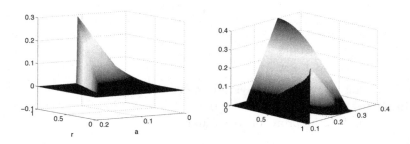

Figure 5.   Left: $\Delta_{12}(\alpha)$ when $\Delta(\alpha) = 0$. Right:$\Delta_{12}(\alpha) - \Delta(\alpha)$ when $\Delta(\alpha) > 0$

6. Conclusions

We have considered the effects on quantum correlations that may arise from
the coupling of a bipartite system to suitably engineered environments. We
have showed that entanglement between two qubits can be generated by
purely dissipative means, can persist asymptotically and that, contrary to

expectations, it can be produced at a rate larger than the entropy production. Passing from two to three qubits, we have dicussed an interesting scenario where appending a totally unpolarized qubit can improve on the entanglement of two qubits when they evolve in a weakly and symmetrically coupled environment. These analytical results have been obtained by means of the available tools concerning the stationary states of quantum dynamical semigroups.

Bibliography

1. M.A. Nielsen, I.L. Chuang, *Quantum Computation and Quantum Information*, (Cambridge University Press, Cambridge, 2000).
2. D. Bruss, G. Leuchs, *Lectures on quantum information*, (Wiley-VCH 2007).
3. M. Horodecki, P. Horodecki, R. Horodecki, *Rev. Mod. Phys.* **81**, 865 (2009).
4. R. Alicki, K. Lendi, *Quantum Dynamical Semigroups and Applications*, Lect. Notes Phys. **717**, (Springer-Verlag, Berlin, 2007).
5. H.-P. Breuer, F. Petruccione, *The Theory of Open Quantum Systems* (Oxford University Press, Oxford, 2002).
6. D. Braun, *Phys. Rev. Lett.* **89**, 277901 (2002).
7. A. Beige et al., *J. Mod. Opt.* **47**, 2583 (2000).
8. L. Jakobczyk, *J. Phys. A* **35**, 6383 (2002).
9. L. Jakobczyk, *J. Phys. B* **43**, 015502 (2010).
10. F. Benatti, R. Floreanini, M. Piani, *Phys. Rev. Lett.* **91**, 070402 (2003).
11. F. Benatti and R. Floreanini, *Int. J. Mod. Phys. B* **19**, 3063 (2005).
12. A. Isar, *Open Sys. Inf. Dynamics* **16**, 205 (2009).
13. B. Kraus et al., *Phys. Rev. A* **78**, 042307 (2008).
14. V. Vedral, *Journal of Physics: Conference Series* **143**, 012010 (2009).
15. W.K. Wootters, *Phys. Rev. Lett.*, **80**, 2245 (1998).
16. M. Ohya, D. Petz, *Quantum Entropy and Its Use*, (Springer, Berlin 1993).
17. H. Spohn, *Rev. Mod. Phys.* **52**, 569 (1980).
18. M. B. Plenio, V. Vedral, *Cont. Phys.* **39**, 431 (1998).
19. V. Vedral et al., *Phys. Rev. Lett.* **78**, 2275 (1997).
20. V. Vedral, M. B. Plenio, *Phys. Rev. A* **57**, 1619 (1998).
21. F. Benatti, A.M. Liguori, A. Nagy, *J. Math. Phys.* **49**, 042103 (2008).
22. F. Benatti, *Lecture Note Series*, Institute for Mathematical Sciences, National University of Singapore **20**, 133 (2010).
23. F. Benatti, A.M. Liguori, G. Paluzzano, *J. Phys. A* **43**, 045304 (2010).
24. F. Benatti, R. Floreanini, *J. Opt. B* **7**, S429 (2005).
25. A. Frigerio, *Lett. Math. Phys.* **2**, 79 (1977).
26. H. Spohn, *Lett. Math. Phys.* **2**, 33 (1977).
27. A. Frigerio, *Comm. Math. Phys.* **63**, 269 (1978).
28. F. Fagnola, R. Rebolledo, *Lec. Notes in Math.* **1882**, 161 (2006).
29. K. Dietz, *J. Phys. A* **37**, 6143 (2004).
30. B. Baumgartner, H. Narnhofer, W. Thirring, *J. Phys. A* **41**, 065201 (2008).
31. B. Baumgartner, H. Narnhofer, *J. Phys. A* **41**, 395303 (2008).

32. F. Benatti, A. Nagy, *Three Open Qubit Asymptotic Entanglement*, submitted to *J. Math. Phys.* (2010).

Constructing positive maps in matrix algebras

D. Chruściński*

A. Kossakowski**

*Institute of Physics, Nicolaus Copernicus University,
Grudziądzka 5/7, 87-100 Toruń, Poland*
* E-mail: darch@fizyka.umk.pl
** E-mail: kossak@fizyka.umk.pl

We analyze several classes of positive maps in finite dimensional matrix algebras. Such maps provide basic tools for studying quantum entanglement in finite dimensional Hilbert spaces. Instead of presenting the most general approach to the problem we concentrate on specific examples. We stress that there is no general construction of positive maps. We show how to generalize well known maps in low dimensional algebras: Choi map in $M_3(\mathbb{C})$ and Robertson map in $M_4(\mathbb{C})$.

Keywords: matrix algebras; positive maps; entanglement witnesses.

1. Introduction

One of the most important problems of quantum information theory[1,2] is the characterization of mixed states of composed quantum systems. In particular it is of primary importance to test whether a given quantum state exhibits quantum correlation, i.e. whether it is separable or entangled. For low-dimensional systems there exists simple necessary and sufficient condition for separability. The celebrated Peres-Horodecki criterion[3,4] states that a state of a bipartite system living in $\mathbb{C}^2 \otimes \mathbb{C}^2$ or $\mathbb{C}^2 \otimes \mathbb{C}^3$ is separable iff its partial transpose is positive. Unfortunately, for higher-dimensional systems there is no single *universal* separability condition.

It turns out that the above problem may be reformulated in terms of positive linear maps in operator algebras: a state ρ in $\mathcal{H}_1 \otimes \mathcal{H}_2$ is separable iff $(\mathrm{id} \otimes \varphi)\rho$ is positive for any positive map φ which sends positive operators on $\mathcal{H}_2$ into positive operators on $\mathcal{H}_1$. Therefore, a classification of positive linear maps between operator algebras $\mathcal{B}(\mathcal{H}_1)$ and $\mathcal{B}(\mathcal{H}_2)$ is of primary importance. Unfortunately, in spite of the considerable effort, the structure of positive maps is rather poorly understood. The mathematical research was initiated by Størmer i Arverson[5-7] the 60. and continued in the 70. by Choi[8,9] and Woronowicz[10] (see also[6]). An interesting contribution to the structure of positive maps was made also by Robertson[11] who provided new examples of positive maps in $M_4(\mathbb{C})$. Recently, in connection with quantum information problem of positive maps was intensively studied both by mathematicians and physicists (see recent papers,[12,43] see also the monograph by Paulsen[44]). Positive maps play an important role both in physics and mathematics providing generalization of $*$-homomorphisms, Jordan homomorphisms and conditional expectations. Normalized positive maps define affine mappings between sets of states of $\mathbb{C}^*$-algebras.

In this paper we restrict to the finite dimensional case. Clearly, one may discuss much more general approach based on infinite dimensional $\mathbb{C}^*$-algebras.[44] However, the difficult part of the problem is not the most general approach since even for $\mathcal{H}_1 = \mathcal{H}_2 = \mathbb{C}^3$ the problem of constructing positive maps is highly nontrivial. Therefore, instead of studying general case we will work through examples.

2. Preliminaries

In this paper we restrict our analysis to linear maps

$$\Lambda : M_d(\mathbb{C}) \ \rightarrow \ M_d(\mathbb{C}) \ , \tag{1}$$

where $M_d(\mathbb{C})$ denotes a set of $d \times d$ complex matrices. Let $M_d(\mathbb{C})^+$ be a convex set of semi-positive elements in $M_d(\mathbb{C})$, that is, $M_d(\mathbb{C})^+$ defines a space of (unnormalized) states of d-level quantum system. One calls Λ a positive map if $\Lambda a \in M_d(\mathbb{C})^+$ for any $a \in M_d(\mathbb{C})^+$. Similarly, Λ is k-positive if

$$\Lambda_{(k)} := \mathrm{id}_k \otimes \Lambda \ : M_k(\mathbb{C}) \otimes M_d(\mathbb{C}) \ \longrightarrow \ M_k(\mathbb{C}) \otimes M_d(\mathbb{C}) \ , \tag{2}$$

is positive. Finally, Λ is completely positive (CP) if it is k-positive for all k. Let $\mathcal{P}_k$ denotes a convex set of k-positive maps in $M_d(\mathbb{C})$. One has $\mathcal{P}_k \subset \mathcal{P}_l$ for $k > l$. Actually, due to the Choi theorem any d-positive map in $M_d(\mathbb{C})$ is CP, and hence $\mathcal{P}_{\mathrm{CP}} = \mathcal{P}_d$. Therefore, one has the following chain of proper inclusions

$$\mathcal{P}_{\mathrm{CP}} \subset \mathcal{P}_{d-1} \subset \ldots \subset \mathcal{P}_1 \ , \tag{3}$$

where $\mathcal{P}_1$ denotes a set of all positive maps in $M_d(\mathbb{C})$.

Let "T" denotes a transposition map. One calls a linear map Λ k-copositive if the map $\Lambda \circ \mathrm{T}$ is k-positive. Let $\mathcal{P}^k$ denotes a convex set of k-copositive maps. One has

$$\mathcal{P}^{\mathrm{CP}} \subset \mathcal{P}^{d-1} \subset \ldots \subset \mathcal{P}^1 \ , \tag{4}$$

where $\mathcal{P}^1$ denotes a set of all copositive maps in $M_d(\mathbb{C})$, and $\mathcal{P}^{\mathrm{CP}}$ stands for a set of completely copositive maps (CcP). Let $\mathcal{P}_l^k$ denotes a set of maps which are l-positive and k-copositive. One has the following relations

$$\mathcal{P}_{\mathrm{CP}}^{\mathrm{CP}} \subset \mathcal{P}_{d-1}^{d-1} \subset \ldots \subset \mathcal{P}_1^1 \ . \tag{5}$$

A positive map $\Lambda \in \mathcal{P}_1$ is called *decomposable* if

$$\Lambda = \Lambda_1 + \Lambda_2 \ , \tag{6}$$

where $\Lambda_1 \in \mathcal{P}_{\mathrm{CP}}$ and $\Lambda_2 \in \mathcal{P}^{\mathrm{CP}}$. A map which is not decomposable is called *indecomposable*. A positive map $\Lambda \in \mathcal{P}_1$ is called *atomic* if it cannot be written as in (6), where $\Lambda_1 \in \mathcal{P}_2$ and $\Lambda_2 \in \mathcal{P}^2$. It is clear that each atomic map is indecomposable but the converse is not true.

Since $\mathcal{P}_1$ is a convex set it is fully characterised by its extreme elements. Clearly a positive map Λ is *extremal* if for any $\Psi \in \mathcal{P}_1$, a map $\Lambda - \Psi$ is not positive. Finally, a positive map Λ is *optimal* if for any $\Psi \in \mathcal{P}_{\mathrm{CP}}$, a map $\Lambda - \Psi$ is not positive. It is evident that each extremal map is optimal but the converse is not true.

3. The structure of entanglement witnesses

Consider a composite system living in $\mathbb{C}^d \otimes \mathbb{C}^d$. One calls a Hermitian operator $W : \mathbb{C}^d \otimes \mathbb{C}^d \to \mathbb{C}^d \otimes \mathbb{C}^d$ an *entanglement witness* (EW) if

   i) $\langle \psi \otimes \phi | W | \psi \otimes \phi \rangle \geq 0$ for all $\psi \otimes \phi \in \mathbb{C}^d \otimes \mathbb{C}^d$,

   ii) W is not positive, i.e., it has at least one strictly negative eigenvalue. A linear operator in $\mathbb{C}^d \otimes \mathbb{C}^d$ satisfying *i)* is called *block positive*. Hence, EW is a block positive operator which is not positive. There is a simple relation between block positive operators and positive maps:[45] W is block positive if and only if

$$W = (\mathrm{id} \otimes \Lambda) P_d^+ \ , \tag{7}$$

for some positive map Λ in $M_d(\mathbb{C})$. In the above formulae P_d^+ denotes a canonical maximally entangled state in $\mathbb{C}^d \otimes \mathbb{C}^d$

$$P_d^+ = \frac{1}{d} \sum_{i,j=1}^d e_{ij} \otimes e_{ij} \ , \tag{8}$$

where $\{e_1, \ldots, e_d\}$ stands for an orthonormal basis in $\mathbb{C}^d \otimes \mathbb{C}^d$, and $e_{ij} = |e_i\rangle\langle e_j|$ denotes an orthonormal basis in $M_d(\mathbb{C})$ (recall, that $M_d(\mathbb{C})$ is a

d^2–dimensional complex Hilbert space equipped with the Hilbert-Schmidt inner product $(A, B) := \mathrm{Tr}(AB^\dagger))$. In particular W is positive iff Λ is CP.

Now, for any density operator $\rho \in (M_d \otimes M_d)^+$ denote by $\mathrm{SN}(\rho)$ the corresponding Schmidt number of ρ. Let us recall[26] that for any normalized positive operator ρ on $\mathcal{H} \otimes \mathcal{H}$ one may define its Schmidt number

$$\mathrm{SN}(\rho) = \min_{p_k, \psi_k} \left\{ \max_k \mathrm{SR}(\psi_k) \right\} , \tag{9}$$

where the minimum is taken over all possible pure states decompositions

$$\rho = \sum_k p_k |\psi_k\rangle\langle\psi_k| , \tag{10}$$

with $p_k \geq 0$, $\sum_k p_k = 1$ and ψ_k are normalized vectors in $\mathcal{H} \otimes \mathcal{H}$. Let us introduce the following family of positive cones:

$$V_r = \{ \rho \in (M_d \otimes M_d)^+ \mid \mathrm{SN}(\rho) \leq r \} . \tag{11}$$

One has the following chain of inclusions

$$V_1 \subset \ldots \subset V_d \equiv (M_d \otimes M_d)^+ . \tag{12}$$

Clearly, V_1 is a cone of separable (unnormalized) states and $V_d \smallsetminus V_1$ stands for a set of entangled states. Note, that a partial transposition $(\mathrm{id} \otimes \mathrm{T})$ gives rise to another family of cones:

$$V^l = (\mathrm{id} \otimes \mathrm{T})V_l , \tag{13}$$

such that $V^1 \subset \ldots \subset V^d$. One has $V_1 = V^1$, together with the following hierarchy of inclusions

$$V_1 = V_1 \cap V^1 \subset V_2 \cap V^2 \subset \ldots \subset V_d \cap V^d . \tag{14}$$

Note, that $V_d \cap V^d$ is a convex set of PPT (unnormalized) states. Finally, $V_r \cap V^s$ is a convex subset of PPT states ρ such that $\mathrm{SN}(\rho) \leq r$ and $\mathrm{SN}(\rho^\Gamma) \leq s$.

Proposition 3.1.

Let $\Lambda : M_d(\mathbb{C}) \to M_d(\mathbb{C})$ be a linear map. $\Lambda \in \mathcal{P}_k$ if and only if

$$(\mathrm{id} \otimes \Lambda)V_k \subset V_d . \tag{15}$$

$\Lambda \in \mathcal{P}^k$ if and only if

$$(\mathrm{id} \otimes \Lambda)V^k \subset V_d . \tag{16}$$

Finally, $\Lambda \in \mathcal{P}_l^k$ if and only if

$$(\mathrm{id} \otimes \Lambda)V^k \cap V_l \subset V_d . \tag{17}$$

Let us denote by W a space of entanglement witnesses, i.e. a space of non-positive Hermitian operators $W \in M_d \otimes M_d$ such that $\mathrm{Tr}(W\rho) \geq 0$ for all $\rho \in V_1$. Define a family of subsets $W_r \subset M_d \otimes M_d$:

$$W_r = \{ W \in M_d \otimes M_d \mid \mathrm{Tr}(W\rho) \geq 0 \, , \, \rho \in V_r \} \, . \tag{18}$$

One calls an EW W from W_r an r-EW. One has

$$(M_d \otimes M_d)^+ \equiv W_d \subset \ldots \subset W_1 \, . \tag{19}$$

Clearly, $W = W_1 \smallsetminus W_d$. Moreover, for any $k > l$, entanglement witnesses from $W_l \smallsetminus W_k$ can detect entangled states from $V_k \smallsetminus V_l$, i.e. states ρ with Schmidt number $l < \mathrm{SN}(\rho) \leq k$. In particular $W \in W_k \smallsetminus W_{k+1}$ can detect state ρ with $\mathrm{SN}(\rho) = k$.

Consider now the following class of witnesses

$$W_r^s := W_r + (\mathrm{id} \otimes \mathrm{T})W_s \, , \tag{20}$$

that is, $W \in W_r^s$ iff

$$W = P + Q^\Gamma \, , \tag{21}$$

with $P \in W_r$ and $Q \in W_s$. Note, that $\mathrm{Tr}(W\rho) \geq 0$ for all $\rho \in V_r \cap V^s$. Hence such W can detect PPT states ρ such that $\mathrm{SN}(\rho) \geq r$ or $\mathrm{SN}(\rho^\Gamma) \geq s$. Entanglement witnesses from W_d^d are called decomposable.[27] They cannot detect PPT states. One has the following chain of inclusions:

$$W_d^d \subset \ldots \subset W_2^2 \subset W_1^1 \equiv W \, . \tag{22}$$

The 'weakest' entanglement can be detected by elements from $W_1^1 \smallsetminus W_2^2$. We shall call them *atomic entanglement witnesses*.

4. Basic indecomposable maps in low dimensions

It is well known that all positive maps $\Lambda : M_2(\mathbb{C}) \to M_2(\mathbb{C})$, $\Lambda : M_2(\mathbb{C}) \to M_3(\mathbb{C})$ and $\Lambda : M_3(\mathbb{C}) \to M_2(\mathbb{C})$ are decomposable.[10]

4.1. *Choi map in $M_3(\mathbb{C})$*

The first example of an indecomposable positive linear map in $M_3(\mathbb{C})$ was found by Choi.[8] The (normalized) Choi map reads as follows

$$\Phi_{\mathrm{C}}(e_{ii}) = \sum_{i,j=1}^{3} A_{ij}^{\mathrm{C}} e_{jj} \, ,$$

$$\Phi_{\mathrm{C}}(e_{ij}) = -\frac{1}{2} e_{ij} \, , \quad i \neq j \, , \tag{23}$$

where $||A_{ij}^C||$ is the following doubly stochastic matrix:

$$A_{ij}^C = \frac{1}{2} \begin{pmatrix} 1 & 1 & 0 \\ 0 & 1 & 1 \\ 1 & 0 & 1 \end{pmatrix} . \tag{24}$$

This map may be generalized as follows:[14] for any $a, b, c \geq 0$ let us define

$$\Phi[a, b, c](e_{ii}) = \sum_{i,j=1}^{3} A_{ij} e_{jj} ,$$

$$\Phi[a, b, c](e_{ij}) = -\frac{1}{a+b+c} e_{ij} , \quad i \neq j , \tag{25}$$

with

$$A_{ij} = \frac{1}{a+b+c} \begin{pmatrix} a & b & c \\ c & a & b \\ b & c & a \end{pmatrix} . \tag{26}$$

Clearly, $\Phi^C = \Phi[1, 1, 0]$. The map $\Phi[1, 0, \mu]$ with $\mu \geq 1$ is an example of an indecomposable map introduced by Choi.[9] Now, it was shown[14] that $\Phi[a, b, c]$ is a positive map if and only if the following conditions are satisfied:

(i) $0 \leq a < 2$,

(ii) $a + b + c \geq 2$,

(iii) if $a \leq 1$, then $(1 - a)^2 \leq bc$.

Moreover, $\Phi[a, b, c]$ is indecomposable iff additionally

(iv) $bc < (2 - a)^2/4$.

Actually, $\Phi[a, b, c]$ is indecomposable if and only if it is atomic, i.e. it cannot be decomposed into the sum of 2-positive and 2-copositive maps. The corresponding entanglement witness reads as follows

$$W[a, b, c] = \frac{1}{3(a+b+c)} \begin{pmatrix} a & \cdot & \cdot & \cdot & -1 & \cdot & \cdot & \cdot & -1 \\ \cdot & b & \cdot & \cdot & \cdot & \cdot & \cdot & \cdot & \cdot \\ \cdot & \cdot & c & \cdot & \cdot & \cdot & \cdot & \cdot & \cdot \\ \cdot & \cdot & \cdot & c & \cdot & \cdot & \cdot & \cdot & \cdot \\ -1 & \cdot & \cdot & \cdot & a & \cdot & \cdot & \cdot & -1 \\ \cdot & \cdot & \cdot & \cdot & \cdot & b & \cdot & \cdot & \cdot \\ \cdot & \cdot & \cdot & \cdot & \cdot & \cdot & b & \cdot & \cdot \\ \cdot & \cdot & \cdot & \cdot & \cdot & \cdot & \cdot & c & \cdot \\ -1 & \cdot & \cdot & \cdot & -1 & \cdot & \cdot & \cdot & a \end{pmatrix} , \tag{27}$$

where to maintain more transparent form we replace all zeros by dots. Note, that $\Phi[1,0,1] = \Phi_{\mathbb{C}}^{\#}$, where the dual map $\Lambda^{\#}$ is defined by: $\mathrm{Tr}(A\Lambda^{\#}(B)) = \mathrm{Tr}(\Lambda(A)B)$ for any $A, B \in M_3(\mathbb{C})$. Note, that $\Phi[0,1,1]$ reproduces the reduction map in $M_3(\mathbb{C})$.

4.2. Robertson map in $M_4(\mathbb{C})$

Let us observe that $M_{2K}(\mathbb{C}) = M_2(\mathbb{C}) \otimes M_K(\mathbb{C})$ and hence any matrix $X \in M_{2K}(\mathbb{C})$ may be represented as follows

$$X = \sum_{k,l=1}^{2} e_{kl} \otimes X_{kl} \ , \tag{28}$$

where $X_{kl} \in M_K(\mathbb{C})$. In what follows we shall use the following notation

$$X = \left(\begin{array}{c|c} X_{11} & X_{12} \\ \hline X_{21} & X_{22} \end{array}\right) \ , \tag{29}$$

to display the block structure of X. Robertson map[11] in $M_4(\mathbb{C})$ is defined as follows:

$$\Phi_4\left(\begin{array}{c|c} X_{11} & X_{12} \\ \hline X_{21} & X_{22} \end{array}\right) = \frac{1}{2}\left(\begin{array}{c|c} \mathbb{I}_2 \,\mathrm{Tr}X_{22} & -[X_{12} + R_2(X_{21})] \\ \hline -[X_{21} + R_2(X_{12})] & \mathbb{I}_2 \,\mathrm{Tr}X_{11} \end{array}\right) \ , \tag{30}$$

where R_2 is a reduction map in $M_2(\mathbb{C})$

$$R_2(X) = \mathbb{I}_2\mathrm{Tr}X - X \ . \tag{31}$$

The corresponding entanglement witness reads as follows

$$W_R = \frac{1}{8}\left(\begin{array}{cccc|cccc|cccc|cccc}
\cdot & \cdot & \cdot & \cdot & \cdot & \cdot & \cdot & \cdot & \cdot & \cdot & -1 & \cdot & \cdot & \cdot & \cdot & -1 \\
\cdot & \cdot & 1 & \cdot & \cdot & \cdot & \cdot & \cdot & \cdot & \cdot & \cdot & \cdot & \cdot & -1 & \cdot & \cdot \\
\cdot & \cdot & \cdot & 1 & \cdot & \cdot & \cdot & \cdot & \cdot & 1 & \cdot & \cdot & \cdot & \cdot & \cdot & \cdot \\
\cdot & \cdot & \cdot & \cdot & \cdot & \cdot & \cdot & \cdot & \cdot & \cdot & \cdot & \cdot & \cdot & \cdot & \cdot & \cdot \\
\hline
\cdot & \cdot & \cdot & \cdot & \cdot & \cdot & \cdot & \cdot & \cdot & \cdot & -1 & \cdot & \cdot & \cdot & \cdot & -1 \\
\cdot & \cdot & \cdot & \cdot & \cdot & \cdot & 1 & \cdot & \cdot & \cdot & \cdot & \cdot & -1 & \cdot & \cdot & \cdot \\
\cdot & \cdot & \cdot & \cdot & \cdot & \cdot & \cdot & 1 & 1 & \cdot & \cdot & \cdot & \cdot & \cdot & \cdot & \cdot \\
\cdot & \cdot & \cdot & \cdot & \cdot & \cdot & \cdot & 1 & 1 & \cdot & \cdot & \cdot & \cdot & \cdot & \cdot & \cdot \\
\hline
\cdot & \cdot & \cdot & \cdot & \cdot & \cdot & 1 & \cdot & \cdot & 1 & \cdot & \cdot & \cdot & \cdot & \cdot & \cdot \\
\cdot & \cdot & \cdot & -1 & \cdot & \cdot & \cdot & \cdot & \cdot & 1 & \cdot & \cdot & \cdot & \cdot & \cdot & \cdot \\
-1 & \cdot & \cdot & \cdot & -1 & \cdot & \cdot & \cdot & \cdot & \cdot & \cdot & \cdot & \cdot & \cdot & \cdot & \cdot \\
\cdot & \cdot & \cdot & \cdot & \cdot & \cdot & \cdot & \cdot & \cdot & \cdot & \cdot & \cdot & \cdot & \cdot & \cdot & \cdot \\
\hline
\cdot & \cdot & \cdot & \cdot & \cdot & \cdot & -1 & \cdot & \cdot & \cdot & \cdot & \cdot & 1 & \cdot & \cdot & \cdot \\
\cdot & \cdot & \cdot & -1 & \cdot & \cdot & \cdot & \cdot & \cdot & \cdot & \cdot & \cdot & \cdot & 1 & \cdot & \cdot \\
\cdot & \cdot & \cdot & \cdot & \cdot & \cdot & \cdot & \cdot & \cdot & \cdot & \cdot & \cdot & \cdot & \cdot & \cdot & \cdot \\
-1 & \cdot & \cdot & \cdot & -1 & \cdot & \cdot & \cdot & \cdot & \cdot & \cdot & \cdot & \cdot & \cdot & \cdot & \cdot
\end{array}\right) \ . \tag{32}$$

Note, that W_R has single negative eigenvalue '$-1/4$', '0' (with multiplicity 10) and '$+1/4$' (with multiplicity 5).

5. Indecomposable maps in $M_d(\mathbb{C})$ — generalized Choi maps

In this section we provide several examples of positive maps in $M_d(\mathbb{C})$ which generalize Choi map in $M_3(\mathbb{C})$.

Example 5.1.

The Choi map in $M_3(\mathbb{C})$ may be generalized to a positive map in $M_d(\mathbb{C})$ as follows:[18] let S be a unitary shift defined by:

$$S\, e_i = e_{i+1} \, , \qquad i = 1, \ldots, d \, ,$$

where the indices are understood mod d. One defines

$$\tau_{d,k}(X) = (d-k)\, \epsilon(X) + \sum_{i=1}^{k} \epsilon(S^i\, X\, S^{*i}) - X \, , \qquad k = 0, 1, 2, \ldots, d-1 \, , \quad (33)$$

where $\epsilon(X)$ denotes the following projector

$$\epsilon(X) = \sum_{k=1}^{d} e_{kk} X e_{kk} \, .$$

The map $\tau_{d,0}$ defined is completely positive and the map $\tau_{d,d-1}$ reproduces the reduction map in $M_d(\mathbb{C})$ (and hence it is completely copositive). Note that $\tau_{d,k}(\mathbb{I}_d) = (d-1)\mathbb{I}_d$, and $Tr\,\tau_{d,k}(X) = (d-1)\,Tr\,X$, hence the normalized maps

$$\Phi_{d,k}(X) = \frac{1}{d-1}\, \tau_{d,k}(X) \, , \qquad\qquad (34)$$

are doubly stochastic. In particular $\Phi[1,0,1] = \Phi_{3,1}$.

Example 5.2. A class of maps $\varphi_{\mathbf{p}}$ parameterized by $d+1$ parameters $\mathbf{p} = (p_0, p_1, \ldots, p_d)$:

$$\begin{aligned}
\varphi_{\mathbf{p}}(e_{11}) &= p_0 e_{11} + p_d e_{dd} \, , \\
\varphi_{\mathbf{p}}(e_{22}) &= p_0 e_{22} + p_1 e_{11} \, , \\
&\ \ \vdots \\
\varphi_{\mathbf{p}}(e_{dd}) &= p_0 e_{dd} + p_{d-1} e_{d-1,d-1} \, , \\
\varphi_{\mathbf{p}}(e_{ij}) &= -e_{ij} \, , \qquad i \neq j \, .
\end{aligned} \qquad (35)$$

It was shown[13,19] *that if*

$$a) \quad p_1, \ldots, p_d > 0 \ ,$$

$$b) \quad d - 1 > p_0 \geq d - 2 \ ,$$

$$c) \quad p_1 \cdot \ldots \cdot p_d \geq (d - 1 - p_0)^d \ ,$$

then $\varphi_{\mathbf{p}}$ is a positive indecomposable map. Actually, $\varphi_{\mathbf{p}}$ is atomic, i.e. it cannot be decomposed into the sum of a 2-positive and 2-copositive maps. In particular the corresponding EW for $d = 3$ reads as follows

$$W_{p_0,p_1,p_2,p_3} = \frac{1}{3p_0 + p_1 + p_2 + p_3} \begin{pmatrix} p_0 & \cdot & \cdot & \vline & \cdot & -1 & \cdot & \vline & \cdot & \cdot & -1 \\ \cdot & \cdot & \cdot & \vline & \cdot & \cdot & \cdot & \vline & \cdot & \cdot & \cdot \\ \cdot & \cdot & p_3 & \vline & \cdot & \cdot & \cdot & \vline & \cdot & \cdot & \cdot \\ \hline \cdot & \cdot & \cdot & \vline & p_1 & \cdot & \cdot & \vline & \cdot & \cdot & \cdot \\ -1 & \cdot & \cdot & \vline & \cdot & p_0 & \cdot & \vline & \cdot & \cdot & -1 \\ \cdot & \cdot & \cdot & \vline & \cdot & \cdot & \cdot & \vline & \cdot & \cdot & \cdot \\ \hline \cdot & \cdot & \cdot & \vline & \cdot & \cdot & \cdot & \vline & \cdot & p_2 & \cdot \\ \cdot & \cdot & \cdot & \vline & \cdot & \cdot & \cdot & \vline & \cdot & \cdot & \cdot \\ -1 & \cdot & \cdot & \vline & \cdot & -1 & \cdot & \vline & \cdot & \cdot & p_0 \end{pmatrix} . \quad (36)$$

In particular if $p_1 = p_2 = p_3 = c$, then $W_{p_0,p_1,p_2,p_3} = W[p_0, 0, c]$.

Example 5.3. *Consider now a class of doubly stochastic matrices $||A_{ij}||$ satisfying*

$$\sum_{k=1}^{d} A_{ik} A_{jk} = \frac{1}{(d-1)^2} \left(\lambda^2 \delta_{ij} + d - 2 + \frac{1 - \lambda^2}{d} \right) , \quad (37)$$

with $\lambda \in [0, 1]$. Note, that for $d = 2$ and $\lambda = 1$ the 2×2 matrix $||A_{ij}||$ is orthogonal.

One proves the following[34]

Proposition 5.1. *A linear map in $M_d(\mathbb{C})$*

$$\varphi(e_{ii}) = \sum_{j=1}^{d} A_{ij} e_{jj} \ ,$$

$$\varphi(e_{ii}) = -\frac{1}{d-1} e_{ij} \ , \quad i \neq j \ , \quad (38)$$

with $||A_{ij}||$ being a doubly stochastic matrix satisfying (37) is positive.

For example if $d = 3$ and $||A_{ij}||$ is defined by

$$A_{ij} = \begin{pmatrix} \alpha_0 & \alpha_1 & \alpha_2 \\ \alpha_2 & \alpha_0 & \alpha_1 \\ \alpha_1 & \alpha_2 & \alpha_0 \end{pmatrix} , \quad (39)$$

with $\alpha_0 + \alpha_1 + \alpha_2 = 1$, *then conditions (37) imply*

$$\alpha_0^2 + \alpha_1^2 + \alpha_2^2 = \frac{1}{2} \ .$$

Note that taking $\alpha_0 = \alpha_1 = 1/2$ and $\alpha_2 = 0$ one reproduces the Choi map. Interestingly, one has the following result[34]

Theorem 5.1. *Let $\varphi : M_d \longrightarrow M_d$ be a positive map defined by (38) with a doubly stochastic matrix $\|A_{ij}\|$ satisfying (37). Suppose that a matrix $\|A_{ij}\|$ is circulant, i.e.*

$$A_{ij} = \begin{pmatrix} \alpha_0 & \alpha_1 & \alpha_2 & \dots & \alpha_{d-1} \\ \alpha_{d-1} & \alpha_0 & \alpha_1 & \dots & \alpha_{d-2} \\ \vdots & \vdots & \vdots & \ddots & \vdots \\ \alpha_1 & \alpha_2 & \alpha_3 & \dots & \alpha_0 \end{pmatrix} , \qquad (40)$$

with $\alpha_i \geq 0$, and $\alpha_0 + \alpha_1 + \ldots + \alpha_{d-1} = 1$. Then φ is indecomposable if :
i) for $d = 2k + 1$ one of the following two conditions is satisfied

$$\text{1)} \quad \begin{cases} \alpha_1 + \ldots + \alpha_k > 0 \\ \alpha_1 + \ldots + \alpha_k \neq \alpha_{k+1} + \ldots + \alpha_{2k} \end{cases} ,$$

$$\text{2)} \quad \begin{cases} \alpha_1 + \ldots + \alpha_k = 0 \\ 1 > \alpha_0 > 0 \end{cases} ,$$

ii) for $d = 2k$ one of the following two conditions is satisfied

$$\text{1)} \quad \begin{cases} \alpha_1 + \ldots + \alpha_{k-1} > 0 \\ \alpha_1 + \ldots + \alpha_{k-1} \neq \alpha_{k+1} + \ldots + \alpha_{2k-1} \end{cases} ,$$

$$\text{2)} \quad \begin{cases} \alpha_1 + \ldots + \alpha_{k-1} = 0 \\ 1 > \alpha_0 + \alpha_k > 0 \end{cases} .$$

This example shows how to produce 'quantum' positive map in $M_d(\mathbb{C})$ out of 'classical' positive map defined by a doubly stochastic matrix $\|A_{ij}\|$. We stress, that a 'classical' map is necessarily CP but the 'quantum' one is not. Moreover, the 'quantum' counterpart is indecomposable.

6. Positive maps from spectral conditions

Any entanglement witness W can be represented as a difference $W = W_+ - W_-$, where both W_+ and W_- are semi-positive operators in $\mathbb{C}^d \otimes \mathbb{C}^d$. However, there is no general method to recognize that W defined by $W_+ - W_-$ is indeed an EW. One particular method based on spectral properties of W was presented in.[38] Let ψ_α ($\alpha = 1, \ldots, D = d^2$) be an orthonormal basis in $\mathbb{C}^d \otimes \mathbb{C}^d$ and denote by P_α the corresponding projector $P_\alpha = |\psi_\alpha\rangle\langle\psi_\alpha|$. It leads therefore to the following spectral resolution of identity

$$\mathbb{I}_d \otimes \mathbb{I}_d = \sum_{\alpha=1}^{D} P_\alpha \ . \tag{41}$$

Having defined eigenvectors of W one needs the corresponding eigenvalues: let

$$\lambda_\alpha^- \leq 0 \ , \quad \alpha = 1, \ldots, L < D \ ,$$

and

$$\lambda_\alpha^+ > 0 \ , \quad \alpha = L+1, \ldots, D \ ,$$

that is,

$$W_- = -\sum_{\alpha=1}^{L} \lambda_\alpha^- P_\alpha \ , \quad W_+ = \sum_{\alpha=L+1}^{D} \lambda_\alpha^+ P_\alpha \ .$$

Our problem is to find the condition for the spectrum $\{\lambda_\alpha^-, \lambda_\alpha^+\}$ which guarantees that W is block positive. Consider a normalized vector $\psi \in \mathbb{C}^d \otimes \mathbb{C}^d$ and let

$$s_1(\psi) \geq \ldots \geq s_d(\psi) \ ,$$

denote its Schmidt coefficients. For any $1 \leq k \leq d$ one defines k-norm of ψ by the following formula

$$||\psi||_k^2 = \sum_{j=1}^{k} s_j^2(\psi) \ . \tag{42}$$

It is clear that

$$||\psi||_1 \leq ||\psi||_2 \leq \ldots \leq ||\psi||_d \ . \tag{43}$$

Note that $||\psi||_1$ gives the maximal Schmidt coefficient of ψ, whereas due to the normalization, $||\psi||_d^2 = \langle\psi|\psi\rangle = 1$. In particular, if ψ is maximally entangled then

$$||\psi||_k^2 = \frac{k}{d} \ . \tag{44}$$

Equivalently one may define k-norm of ψ by

$$||\psi||_k^2 = \max_\phi |\langle\psi|\phi\rangle|^2 \ , \tag{45}$$

where the maximum runs over all normalized vectors ϕ such that $\mathrm{SR}(\psi) \leq k$ (such ϕ is usually called k-separable). Recall that a Schmidt rank of ψ – $\mathrm{SR}(\psi)$ – is the number of non-vanishing Schmidt coefficients of ψ. One calls entanglement witness W a k-EW if $\langle \psi | W | \psi \rangle \geq 0$ for all ψ such that $\mathrm{SR}(\psi) \leq k$. One has the the following

Theorem 6.1. *Let $\sum_{\alpha=1}^{L} ||\psi_\alpha||_k^2 < 1$. If the following spectral conditions are satisfied*

$$\lambda_\alpha^+ \geq \mu_k \ , \quad \alpha = L+1, \ldots, D \ , \tag{46}$$

where

$$\mu_\ell := \frac{\sum_{\alpha=1}^{L} |\lambda_\alpha^-| \, ||\psi_\alpha||_\ell^2}{1 - \sum_{\alpha=1}^{L} ||\psi_\alpha||_\ell^2} \ , \tag{47}$$

then W is an k-EW. If moreover $\sum_{\alpha=1}^{L} ||\psi_\alpha||_{k+1}^2 < 1$ and

$$\mu_{k+1} > \lambda_\alpha^+ \ , \quad \alpha = L+1, \ldots, D \ , \tag{48}$$

then W being k-EW is not $(k+1)$-EW.

For the proof see Ref.[38]

Remark 6.1. *If P_1 is the maximally entangled state in $\mathbb{C}^d \otimes \mathbb{C}^d$, i.e. $F = U/\sqrt{d}$ with unitary U, then the above theorem reproduces 25 years old result by Takasaki and Tomiyama.[24] For $k = 1$ and arbitrary P_1 the formula $\lambda_\alpha^+ \geq \mu_1$ ($\alpha = 2, \ldots, d^2$) was derived by Benatti et. al.[25]*

Interestingly, one has the following

Theorem 6.2. $W = W_+ - W_-$ *is a decomposable EW.*

The proof is easy[39]: note that $W = A + B$, where

$$A = \sum_{\alpha=L+1}^{D} (\lambda_\alpha^+ - \mu_1) P_\alpha \ , \tag{49}$$

and

$$B = \mu_1 \mathbb{I}_d \otimes \mathbb{I}_d - \sum_{\alpha=1}^{L} (|\lambda_\alpha^-| + \mu_1) P_\alpha \ . \tag{50}$$

Now, since $\lambda_\alpha^+ \geq \mu_1$, for $\alpha = L+1, \ldots, D$, it is clear that $A \geq 0$. The partial transposition of B reads as follows

$$B^\Gamma = \mu_1 \mathbb{I}_d \otimes \mathbb{I}_d - \sum_{\alpha=1}^{L} (|\lambda_\alpha^-| + \mu_1) P_\alpha^\Gamma \ . \tag{51}$$

Let us recall that the spectrum of the partial transposition of rank-1 projector $|\psi\rangle\langle\psi|$ is well know: the nonvanishing eigenvalues of $|\psi\rangle\langle\psi|^\Gamma$ are given by $s_\alpha^2(\psi)$ and $\pm s_\alpha(\psi)s_\beta(\psi)$, where $s_1(\psi) \geq \ldots \geq s_d(\psi)$ are Schmidt coefficients of ψ. Therefore, the smallest eigenvalue of B^Γ (call it $b_{\min}$) satisfies

$$b_{\min} \geq \mu_1 - \sum_{\alpha=1}^{L}(|\lambda_\alpha^-| + \mu_1)\|\psi_\alpha\|_1^2 , \tag{52}$$

and using the definition of μ_1 (cf. Eq. (47)) one gets

$$b_{\min} \geq 0 , \tag{53}$$

which implies $B^\Gamma \geq 0$. Hence, the entanglement witness W is decomposable.

Remark 6.2. *Interestingly, saturating the bound (46), i.e. taking*

$$\lambda_\alpha^+ = \mu_1 , \quad \alpha = L+1,\ldots,D , \tag{54}$$

one has $A = 0$ and hence $W = Q^\Gamma$ with $Q = B^\Gamma \geq 0$ which shows that the corresponding positive map $\Lambda : M_d(\mathbb{C}) \to M_d(\mathbb{C})$ defined by

$$\Lambda(X) = \mathrm{Tr}_1(W \cdot X^T \otimes \mathbb{I}_d) , \tag{55}$$

is completely co-positive, i.e. $\Lambda \circ T$ is completely positive. Note that

$$\Lambda(X) = \mu_1 \mathbb{I}_d \mathrm{Tr}\, X - \sum_{\alpha=1}^{L}(\mu_1 + |\lambda_\alpha|)F_\alpha X F_\alpha^\dagger , \tag{56}$$

where F_α is a linear operator $F_\alpha : \mathbb{C}^d \to \mathbb{C}^d$ defined by

$$\psi_\alpha = \sum_{i=1}^{d} e_i \otimes F_\alpha e_i , \tag{57}$$

and $\{e_1,\ldots,e_d\}$ denotes an orthonormal basis in $\mathbb{C}^d$. In particular, if $L = 1$, i.e. there is only one negative eigenvalue, then formula (56) (up to trivial rescaling) gives

$$\Lambda(X) = \kappa\, \mathbb{I}_d \mathrm{Tr}\, X - F_1 X F_1^\dagger , \tag{58}$$

with

$$\kappa = \frac{\mu_1}{\mu_1 + |\lambda_1|} = \|\psi_1\|_1^2 . \tag{59}$$

It reproduces a positive map (or equivalently an EW $W = \kappa\,\mathbb{I}_d \otimes \mathbb{I}_d - P_1$) which is known to be completely co-positive.[29,35,39] If ψ_1 is maximally entangled, that is, $F_1 = U/\sqrt{d}$ for some unitary $U \in U(d)$, then one finds for $\kappa = 1/d$ and the map (58) is unitary equivalent to the reduction map $\Lambda(X) = U R(X) U^\dagger$, where $R_d(X) = \mathbb{I}_d \mathrm{Tr} X - X$.

Example 6.1. *Consider an EW corresponding to the flip operator in $d = 2$:*

$$W = \begin{pmatrix} 1 & \cdot & | & \cdot & \cdot \\ \cdot & \cdot & | & 1 & \cdot \\ \hline \cdot & 1 & | & \cdot & \cdot \\ \cdot & \cdot & | & \cdot & 1 \end{pmatrix} . \tag{60}$$

Its spectral decomposition has the following form: $W_- = |\lambda_1^-|P_1$

$$-\lambda_1^- = \lambda_2^+ = \lambda_3^+ = \lambda_4^+ = 1 ,$$

and

$$\psi_1 = \frac{1}{\sqrt{2}}(|12\rangle - |21\rangle) ,$$

$$\psi_2 = \frac{1}{\sqrt{2}}(|12\rangle + |21\rangle) ,$$

$$\psi_3 = |11\rangle , \quad \psi_4 = |22\rangle .$$

One finds $\mu_1 = 1$ and hence condition (46) is trivially satisfied $\lambda_\alpha^+ \geq \mu_1$ for $\alpha = 2, 3, 4$. We stress that our construction does not recover flip operator in $d > 2$. It has $d(d-1)/2$ negative eigenvalues. Our construction leads to at most $d - 1$ negative eigenvalues.

Example 6.2. *Entanglement witness corresponding to the reduction map:*

$$\lambda_1^- = 1 - d, \quad \lambda_2^+ = \ldots = \lambda_D^+ = 1 ,$$

and

$$W_- = (d - 1)P_d^+ , \quad W_+ = \mathbb{I}_d \otimes \mathbb{I}_d - P_d^+ , \tag{61}$$

where P_d^+ denotes maximally entangled state in $\mathbb{C}^d \otimes \mathbb{C}^d$. Again, one finds $\mu_1 = 1$ and hence condition (46) is trivially satisfied $\lambda_\alpha^+ \geq \mu_1$ for $\alpha = 2, \ldots, D$. Now, since ψ_1 corresponds to the maximally entangled state one has $1 - \|\psi_1\|_2^2 = (d-2)/d < 1$. Hence, condition (48)

$$\mu_2 = 2\frac{d-2}{d-1} > \lambda_\alpha^+ , \quad \alpha = 2, \ldots, D , \tag{62}$$

implies that W is not a 2-EW.

Example 6.3. *A family $W[a, b, c]$ (see (36)). The spectral properties of $W[a, b, c] = W_+ - W_-$ read as follows: $W_- = -\lambda_1^- P_3^+$ and*

$$\lambda_1^- = a - 2 , \quad \lambda_2^+ = \lambda_3^+ = a + 1 ,$$

$$\lambda_4^+ = \lambda_5^+ = \lambda_6^+ = b , \quad \lambda_7^+ = \lambda_8^+ = \lambda_9^+ = c .$$

One finds $\mu_1 = (2-a)/2$ and hence condition (46) implies

$$a \geq 0 , \quad b, c \geq \frac{2-a}{2} . \tag{63}$$

It gives therefore

$$a + b + c \geq 2 , \tag{64}$$

and one easily shows that the conditions 3 is also satisfied. Summarizing: $W[a, b, c]$ belongs to our spectral class if and only if

(1) $0 \leq a < 2$,
(2) $b, c \geq (2-a)/2$.

Note that the Choi witness $W[1, 1, 0]$ does not belong to this class. It is clear since $W[1, 1, 0]$ is indecomposable. It was shown[14] that $W[a, b, c]$ is decomposable if and only if $a \geq 0$ and

$$bc \geq \frac{(2-a)^2}{4} . \tag{65}$$

Hence $W[a, b, c]$ from our spectral class is always decomposable. In particular $W[0, 1, 1]$ reproduces the EW corresponding to the reduction map in $d = 3$. Note, that there are entanglement witnesses $W[a, b, c]$ which are decomposable, i.e. satisfy (65), but do not belong to or spectral class. Similarly one can check when $W[a, b, c]$ defines 2-EW. One finds $\mu_2 = 2(2-a)$ and hence condition (46) implies

(1) $1 \leq a < 2$,
(2) $b, c \geq 2(2-a)$.

Clearly, any 2-EW from our class is necessarily decomposable. It was shown[14] that all 2-EW from the class $W[a, b, c]$ are decomposable.

7. Bell-diagonal entanglement witnesses

Let us define a generalized Bell states[46] in $\mathbb{C}^d \otimes \mathbb{C}^d$

$$\psi_{mn} = (\mathbb{I}_d \otimes U_{mn})\psi_d^+ , \tag{66}$$

where U_{mn} are unitary matrices defined as follows

$$U_{mn}e_k = \lambda^{mk}e_{k+n} , \tag{67}$$

with

$$\lambda = e^{2\pi i/d} . \tag{68}$$

The matrices U_{mn} define an orthonormal basis in the space $M_d(\mathbb{C})$ of complex $d \times d$ matrices. One easily shows

$$\mathrm{Tr}(U_{mn}U_{rs}^\dagger) = d\,\delta_{mr}\delta_{ns} \ . \tag{69}$$

Some authors[47] call U_{mn} generalized spin matrices since for $d = 2$ they reproduce standard Pauli matrices:

$$U_{00} = \mathbb{I}_2 \ , \ U_{01} = \sigma_1 \ , \ U_{10} = i\sigma_2 \ , \ U_{11} = \sigma_3 \ . \tag{70}$$

One calls a Hermitian operator W in $M_d(\mathbb{C}) \otimes M_d(\mathbb{C})$ *Bell diagonal* if

$$W = \sum_{m,n=0}^{d-1} p_{mn}P_{mn} \ , \tag{71}$$

with $p_{mn} \in \mathbb{R}$, and

$$P_{mn} = |\psi_{mn}\rangle\langle\psi_{mn}| \ . \tag{72}$$

Example 7.1. *Consider the flip operator in $d = 2$. One has*

$$F = P_{00} + P_{10} + P_{01} - P_{11} \ , \tag{73}$$

which proves that F is Bell diagonal and possesses single negative eigenvalue.

Example 7.2. Consider a family $W[a,b,c]$. One finds the following spectral representation

$$W[a,b,c] = (a-2)P_{00} + (a+1)(P_{10} + P_{20}) + b\Pi_1 + c\Pi_2 \ , \tag{74}$$

where

$$\Pi_m = P_{0m} + P_{1m} + P_{2m} \ , \tag{75}$$

which shows that $W[a,b,c]$ is Bell diagonal with a single negative eigenvalue '$a - 2$'.

Example 7.3. *Entanglement witness corresponding to the reduction map $\Lambda(X) = \mathbb{I}\mathrm{Tr}X - X$ in $M_d(\mathbb{C})$. One has*

$$W = \frac{1}{d}\mathbb{I}_d \otimes \mathbb{I}_d - P_d^+ = \frac{1}{d}\sum_{k,l=0}^{d-1} P_{kl} - P_{00} \ , \tag{76}$$

which shows that W is Bell diagonal with a single negative eigenvalue $(1 - d)/d$.

Corollary 7.1. *If $L < d$ and*

$$\lambda_\alpha^+ \geq \mu_1 \ , \quad \alpha = L, \ldots, d^2 - 1 \ , \tag{77}$$

with $\mu_1 = \frac{1}{d-L} \sum_{\alpha=0}^{L-1} |\lambda_\alpha^-|$, *then* $W = W_+ - W_-$ *defines Bell diagonal entanglement witness.*

8. Indecomposable maps in $M_{2k}(\mathbb{C})$

Following[31] and[30] (see[33] for the connection between Robertson map and Breuer-Hall maps) one defines

$$\varphi_{2k}^U(X) = \frac{1}{2(k-1)} \left[R_n(X) - U X^T U^\dagger \right] \ , \tag{78}$$

where U is an antisymmetric unitary matrix in $M_{2k}(\mathbb{C})$. The above normalization guaranties that φ_{2k}^U is unital. The characteristic feature of these maps is that for any rank-1 projector P its image under φ_{2k}^U reads as follows:

$$\varphi_{2k}^U(P) = \frac{1}{2(k-1)} \left[\mathbb{I}_{2k} - P - Q \right] \ , \tag{79}$$

where $Q = U P^T U^\dagger$ is a rank-1 projector orthogonal to P. Hence $\varphi_{2k}^U(P)$ is a projector which proves positivity of φ_{2k}^U. Denote by U_0 the following "canonical" antisymmetric unitary matrix in $M_{2k}(\mathbb{C})$

$$U_0 = \mathbb{I}_k \otimes J \ , \tag{80}$$

where J is a symplectic matrix in $M_2(\mathbb{R})$, that is,

$$J = \begin{pmatrix} 0 & 1 \\ -1 & 0 \end{pmatrix} \ . \tag{81}$$

Note, that if $V \in M_{2k}(\mathbb{R})$ is orthogonal then

$$U = V U_0 V^T \ , \tag{82}$$

is antisymmetric and unitary. Interestingly, the map φ_{2k}^0 corresponding to $U = U_0$ has the following block structure

$$\varphi_{2k}^0 \begin{pmatrix} X_{11} & X_{12} & \cdots & X_{1k} \\ X_{21} & X_{22} & \cdots & X_{2k} \\ \vdots & \vdots & \ddots & \vdots \\ X_{k1} & X_{k2} & \cdots & X_{kk} \end{pmatrix} = \tag{83}$$

$$\frac{1}{2(k-1)} \begin{pmatrix} \mathbb{I}_2(\mathrm{Tr}X - \mathrm{Tr}X_{11}) & -(X_{12}+R_2(X_{21})) & \cdots & -(X_{1k}+R_2(X_{k1})) \\ -(X_{21}+R_2(X_{12})) & \mathbb{I}_2(\mathrm{Tr}X - \mathrm{Tr}X_{22}) & \cdots & -(X_{2k}+R_2(X_{k2})) \\ \vdots & \vdots & \ddots & \vdots \\ -(X_{k1}+R_2(X_{1k})) & -(X_{k2}+R_2(X_{2k})) & \cdots & \mathbb{I}_2(\mathrm{Tr}X - \mathrm{Tr}X_{kk}) \end{pmatrix} \ ,$$

and hence it reduces for $k = 2$ to the Robertson map (30).

In the recent paper[40] we proposed another construction of maps in $M_{2k}(\mathbb{C})$. Now, instead of treating a $2k \times 2k$ matrix X as a $k \times k$ matrix with 2×2 blocks X_{ij} we consider alternative possibility, i.e. we consider X as a 2×2 with $k \times k$ blocks and define

$$\psi_{2k}\left(\begin{array}{c|c} X_{11} & X_{12} \\ \hline X_{21} & X_{22} \end{array}\right) = \frac{1}{k}\left(\begin{array}{c|c} \mathbb{I}_k \operatorname{Tr} X_{22} & -[X_{12} + R_k(X_{21})] \\ \hline -[X_{21} + R_k(X_{12})] & \mathbb{I}_k \operatorname{Tr} X_{11} \end{array}\right) . \quad (84)$$

Again, normalization factor guaranties that the map is unital, i.e. $\psi_{2k}(\mathbb{I}_2 \otimes \mathbb{I}_k) = \mathbb{I}_2 \otimes \mathbb{I}_k$. It is clear that for $k = 2$ one has

$$\psi_4 = \varphi_4^0 . \quad (85)$$

Theorem 8.1. Ψ_{2k} *defines a linear positive map in* $M_{2k}(\mathbb{C})$.

Proof: it is enough to show that each rank-1 projector P is mapped via ψ_{2k} into a positive element in $M_{2k}(\mathbb{C})$, that is, $\psi_{2k}(P) \geq 0$. Let $P = |\psi\rangle\langle\psi|$ with arbitrary ψ from $\mathbb{C}^{2k}$. Now, due to $\mathbb{C}^{2k} = \mathbb{C}^k \oplus \mathbb{C}^k$ one has

$$\psi = \psi_1 \oplus \psi_2 , \quad (86)$$

with $\psi_1, \psi_2 \in \mathbb{C}^k$ and hence

$$P = \left(\begin{array}{c|c} X_{11} & X_{12} \\ \hline X_{21} & X_{22} \end{array}\right) = \left(\begin{array}{c|c} |\psi_1\rangle\langle\psi_1| & |\psi_1\rangle\langle\psi_2| \\ \hline |\psi_2\rangle\langle\psi_1| & |\psi_2\rangle\langle\psi_2| \end{array}\right) . \quad (87)$$

One has therefore

$$\Psi_{2k}(P) = \frac{1}{k}\left(\begin{array}{c|c} \mathbb{I}_k \operatorname{Tr} X_{22} & -A \\ \hline -A^\dagger & \mathbb{I}_k \operatorname{Tr} X_{11} \end{array}\right) , \quad (88)$$

where the linear operator $A : \mathbb{C}^k \to \mathbb{C}^k$ reads as follows

$$A = |\psi_1\rangle\langle\psi_2| - |\psi_2\rangle\langle\psi_1| + \langle\psi_1|\psi_2\rangle \mathbb{I}_k . \quad (89)$$

Let $\langle\psi_j|\psi_j\rangle = a_j^2 > 0$ (if one of a_j vanishes then evidently one has $\Psi_{2k}^0(P) \geq 0$). Defining

$$L = \sqrt{k}\left(\begin{array}{c|c} \mathbb{I}_k \, a_2^{-1} & \mathbb{O}_k \\ \hline \mathbb{O}_k & \mathbb{I}_k \, a_1^{-1} \end{array}\right) , \quad (90)$$

one finds

$$L\Psi_{2k}(P)L^\dagger = \left(\begin{array}{c|c} \mathbb{I}_k & -\widetilde{A} \\ \hline -\widetilde{A}^\dagger & \mathbb{I}_k \end{array}\right) , \quad (91)$$

with

$$\widetilde{A} = |\widetilde{\psi}_1\rangle\langle\widetilde{\psi}_2| - |\widetilde{\psi}_2\rangle\langle\widetilde{\psi}_1| + \langle\widetilde{\psi}_1|\widetilde{\psi}_2\rangle \mathbb{I}_k , \quad (92)$$

and normalized $\widetilde{\psi}_j = \psi_j / a_j$. Hence, to show that $\Psi_{2k}(P) \geq 0$ one needs to prove

$$\left(\begin{array}{c|c} \mathbb{I}_k & -\widetilde{A} \\ \hline -\widetilde{A}^\dagger & \mathbb{I}_k \end{array}\right) \geq 0 , \tag{93}$$

for arbitrary $\psi_j \neq 0$. Now, the above condition is equivalent to

$$\widetilde{A}\widetilde{A}^\dagger \leq \mathbb{I}_k . \tag{94}$$

Vectors $\{\psi_1, \psi_2\}$ span 2-dimensional subspace in $\mathbb{C}^k$ and let $\{e_1, e_2\}$ be a 2-dim. orthonormal basis such that $\psi_1 = e_1$ and

$$\psi_2 = e^{i\lambda} s e_1 + c e_2 , \tag{95}$$

with $s = \sin\alpha$, $c = \cos\alpha$ for some angle α. Now, completing the basis $\{e_1, e_2, e_3, \ldots, e_k\}$ in $\mathbb{C}^k$ one easily finds that the matrix elements of $\widetilde{A}$ has a form of the following direct sum

$$\widetilde{A} = \left(\begin{array}{c|c} e^{-i\lambda} s & c \\ \hline -c & e^{i\lambda} s \end{array}\right) \oplus e^{-i\lambda} s \mathbb{I}_{k-2} . \tag{96}$$

Hence

$$\widetilde{A}\widetilde{A}^\dagger = \mathbb{I}_2 \oplus s^2 \mathbb{I}_{k-2} , \tag{97}$$

which proves (94) since all eigenvalues of $\widetilde{A}\widetilde{A}^\dagger$ – $\{1, 1, s^2, \ldots, s^2\}$ – are bounded by 1. $\qquad\square$

Proposition 8.1. Ψ_{2k} *defines an atomic positive map.*

Proof: in order to prove that W_{2k} is atomic one has to define a PPT state D_{2k} such that Schmidt rank of D_{2k} and of its partial transposition D_{2k}^Γ is bounded by 2 and show that $\mathrm{Tr}(W_{2k} D_{2k}) < 0$. Let us introduce the following family of product vectors

$$\phi_1 = e_1 \otimes e_1 , \quad \phi_2 = e_1 \otimes e_{k+1} , \quad \phi_3 = e_k \otimes e_1 , \quad \phi_4 = e_k \otimes e_{2k} ,$$

and

$$\phi_5 = e_{k+1} \otimes e_1 , \quad \phi_6 = e_{k+1} \otimes e_{k+1} , \quad \phi_7 = e_{k+1} \otimes e_{2k} .$$

Define now the following positive operator

$$D_{2k} = \frac{1}{7}\Big(|\phi_1 + \phi_6\rangle\langle\phi_1 + \phi_6| + |\phi_5 - \phi_4\rangle\langle\phi_5 - \phi_4|$$

$$+ |\phi_2\rangle\langle\phi_2| + |\phi_3\rangle\langle\phi_3| + |\phi_7\rangle\langle\phi_7| \Big) . \tag{98}$$

One easily finds for its partial transposition

$$D_{2k}^{\Gamma} = \frac{1}{7}\Big(|\phi_2 + \phi_5\rangle\langle\phi_2 + \phi_5| + |\phi_3 - \phi_7\rangle\langle\phi_3 - \phi_7|$$
$$+ |\phi_1\rangle\langle\phi_1| + |\phi_4\rangle\langle\phi_4| + |\phi_6\rangle\langle\phi_6|\Big). \tag{99}$$

Now, it is clear from that both D_{2k} and D_{2k}^{Γ} are constructed out of rank-1 projectors and Schmidt rank of each projector is 1 or 2. Therefore

$$\mathrm{SN}(D_{2k}) \le 2, \quad \mathrm{SN}(D_{2k}^{\Gamma}) \le 2.$$

Finally, one finds for the trace

$$\mathrm{Tr}(W_{2k}D_{2k}) = -\frac{1}{7k}, \tag{100}$$

which shows that W_{2k} defines atomic entanglement witness. $\square$

9. Conclusions

We analyzed several classes of positive maps in $M_d(\mathbb{C})$. In particular we show how to generalize well known indecomposable maps in $M_3(\mathbb{C})$ (Choi map) and $M_4(\mathbb{C})$ (Robertson map) for higher dimensions. Interestingly, we are able to construct new classes of indecomposable positive maps in $M_d(\mathbb{C})$ for arbitrary $d \ge 3$. We stress that the general problem of constructing positive maps in $M_d(\mathbb{C})$ (even for $d = 3$) is still open.

Acknowledgments

This work was partially supported by the Polish Ministry of Science and Higher Education Grant No 3004/B/H03/2007/33.

Bibliography

1. M. A. Nielsen and I. L. Chuang, *Quantum computation and quantum information*, Cambridge University Press, Cambridge, 2000.
2. R. Horodecki, P. Horodecki, M. Horodecki and K. Horodecki, Rev. Mod. Phys. **81**, 865 (2009).
3. A. Peres, Phys. Rev. Lett. **77**, 1413 (1996).
4. P. Horodecki, Phys. Lett. A **232**, 333 (1997).
5. E. Størmer, Acta Math. **110**, 233 (1963); Trans. Amer. Math. Soc. **120**, 438 (1965); Proc. Am. Math. Soc. **86**, 402 (1982).
6. E. Størmer, in Lecture Notes in Physics **29**, Springer Verlag, Berlin, 1974, pp. 85-106.
7. W. Arverson, Acta Math. **123**, 141 (1969).
8. M.-D. Choi, Lin. Alg. Appl. **10**, 285 (1975); *ibid* **12**, 95 (1975).

9. M.-D. Choi, J. Operator Theory, **4**, 271 (1980).
10. S.L. Woronowicz, Rep. Math. Phys. **10**, 165 (1976); Comm. Math. Phys. **51**, 243 (1976).
11. A.G. Robertson, J. London Math. Soc. (2) **32**, 133 (1985).
12. H. Osaka, Lin. Alg. Appl. **153**, 73 (1991); *ibid* **186**, 45 (1993).
13. H. Osaka, Publ. RIMS Kyoto Univ. **28**, 747 (1992).
14. S. J. Cho, S.-H. Kye, and S.G. Lee, Lin. Alg. Appl. **171**, 213 (1992).
15. H.-J. Kim and S.-H. Kye, Bull. London Math. Soc. **26**, 575 (1994).
16. S.-H. Kye, Math. Proc. Cambridge Philos. Soc. **122**, 45 (1997); Linear Alg. Appl. **362**, 57 (2003).
17. M.-H. Eom and S.-H. Kye, Math. Scand. **86**, 130 (2000).
18. K.-C. Ha, Publ. RIMS, Kyoto Univ., **34**, 591 (1998).
19. K.-C. Ha, Lin. Alg. Appl. **348**, 105 (2002); *ibid* **359**, 277 (2003).
20. K.-C. Ha, S.-H. Kye and Y. S. Park, Phys. Lett. A **313**, 163 (2003); Phys. Lett. A **325**, 315 (2004); J. Phys. A: Math. Gen. **38**, 9039 (2005).
21. B. M. Terhal, Lin. Alg. Appl. **323**, 61 (2001).
22. W.A. Majewski and M. Marcinek, J. Phys. A: Math. Gen. **34**, 5836 (2001).
23. A. Kossakowski, Open Sys. Information Dyn. **10**, 1 (2003).
24. K. Takasaki and J. Tomiyama, Mathematische Zeitschrift **184**, 101 (1983).
25. F. Benatti, R. Floreanini and M. Piani, Open Systems and Inf. Dynamics, **11**, 325-338 (2004).
26. B. Terhal and P. Horodecki, Phys. Rev. A **61**, 040301 (2000); A. Sanpera, D. Bruss and M. Lewenstein, Phys. Rev. A **63**, 050301(R) (2001).
27. M. Lewenstein, B. Kraus, J. I. Cirac and P. Horodecki, Phys. Rev. A **62**, 052310 (2000).
28. A. Sanpera, D. Bruss and M. Lewenstein, Phys. Rev. A **63**, 050301 (2001).
29. O. Gühne and G. Tóth, Phys. Rep. **474**, 1 (2009).
30. W. Hall, J. Phys. A: Math. Gen. **39**, (2006) 14119.
31. H.-P. Breuer, Phys. Rev. Lett. **97**, 0805001 (2006).
32. R. Augusiak and J. Stasińska, New Journal of Physics **11**, 053018 (2009).
33. D. Chruściński and A. Kossakowski, J. Phys. A: Math. Theor. **41** (2008) 215201.
34. D. Chruściński and A. Kossakowski, Open Systems and Inf. Dynamics, **14**, 275 (2007).
35. M. Piani and C. Mora, Phys. Rev. A **75**, 012305 (2007).
36. D. Chruściński and A. Kossakowski, Phys. Lett. A **373** (2009) 2301-2305.
37. D. Chruściński and A. Kossakowski, J. Phys. A: Math. Theor. **41** (2008) 145301.
38. D. Chruściński and A. Kossakowski, Comm. Math. Phys. **290**, 1051 (2009).
39. D. Chruściński, A. Kossakowski and G. Sarbicki, Phys. Rev. A **80** (2009) 042314.
40. D. Chruściński, J. Pytel and G. Sarbicki, Phys. Rev. A **80** (2009) 062314.
41. E. Størmer, J. Funct. Anal. **254**, 2303 (2008); E. Størmer, *Separable states and positive maps II*, arXiv:0803.4417.
42. Ł. Skowronek and K. Życzkowski, J. Phys. A: Math. Theor. **42**, (2009) 325302.

43. Ł. Skowronek and K. Życzkowski, J. Math. Phys. **50**, 062106 (2009).
44. V. Paulsen, *Completely Bounded Maps and Operator Algebras*, Cambridge University Press, 2003.
45. A. Jamiołkowski, Rep. Math. Phys. **3**, 275 (1972).
46. B. Baumgartner, B. Hiesmayr, and H. Narnhofer, Phys. Rev. A **74**, 032327 (2006); J. Phys. A: Math. Theor. **40**, 7919 (2007); Phys. Lett. A **372**, 2190 (2008).
47. O. Pittenger, M.H. Rubin, Lin. Alg. Appl. **390**, 255 (2004).

Quantum processes

M. Fannes*

J. Wouters**

Instituut voor Theoretische Fysica, K.U.Leuven,
3001-Heverlee, Belgium
* E-mail: mark.fannes@fys.kuleuven.be
** E-mail: jeroen.wouters@fys.kuleuven.be

Keywords: quantum process, quasi-free Fermionic process, entropy

1. Introduction

The aim of this note is to present a number of ideas and questions related to the construction of quantum processes. For technical convenience we restrict ourselves to systems with a finite configuration space in a classical context or with d accessible levels in the quantum.

A reversible dynamics of such an isolated system is rather boring: a finite classical configuration space only supports jumps in discrete time while the evolution of a finite level quantum systems is almost periodic in time

$$A_t = \sum_\omega e^{i\omega t} \hat{A}(\omega).$$

Here, the summation runs over the Bohr frequencies of the system, i.e., the spacings between the energy levels of the Hamiltonian and the $\hat{A}(\omega)$ are the Fourier coefficients of the observable A.

The evolution of systems weakly perturbed by an environment can be reasonably described by a stochastic dynamics, even by a stochastic map Γ if we only observe the system at regular time intervals. Such maps will typically turn pure states into mixed states, a clear signature of their randomizing character. In the classical context we deal with matrices of transition probabilities and in the quantum setting with completely positive trace preserving maps. Repeated action of the map on the states yields a dynamics $\{\Gamma^n \mid n \in \mathbb{N}\}$, Markovian in time. A generic Γ has a unique invariant state

ρ_0 into which any initial state is driven

$$\Gamma^n(\text{initial state}) \overset{n \to \infty}{\longrightarrow} \rho_0.$$

Several figures of merit can be defined for this kind of evolution. Two well-known ones are the spectral gap γ of Γ that controls the asymptotic rate of convergence towards the invariant state

$$\|\Gamma^n(\rho) - \rho_0\| \sim (1 - \gamma)^n, \quad n \text{ large}$$

and the minimal output entropy

$$\mathsf{H}^{\min}(\Gamma) := \min\Big(\{\mathsf{H}(\Gamma(\rho)) \mid \rho \text{ state}\}\Big).$$

Here H denotes either the Shannon entropy of a probability vector or the von Neumann entropy of a density matrix.

A stochastic process in discrete time is a different object: it specifies joint probabilities at different times. More precisely such a process is a state ω on a half-chain $\mathbb{N}$. At each site of this chain sits a copy of our classical configuration space or of the quantum d-level system. We assume, moreover, that the chain is stationary: the state ω is invariant under a right shift.

In the classical case the process models a source that is emitting every time unit a letter belonging to the alphabet $\{\epsilon_1, \epsilon_2, \ldots, \epsilon_d\}$. The state ω specifies the probability that the source emits a given word. Let $(\epsilon(0), \epsilon(1), \epsilon(2), \ldots)$ be a random letter string emitted by the source at times $t = 0, 1, 2, \ldots$ then

$$\text{Prob}\Big(\epsilon(0) = \epsilon_0 \ \& \ \epsilon(1) = \epsilon_1 \ \& \ \ldots \ \& \ \epsilon(n) = \epsilon_n\Big) = \omega(\epsilon_0, \epsilon_1, \ldots, \epsilon_n).$$

In the quantum context, the restriction of ω to the first n sites of the chain is a density matrix on $\otimes^n \mathcal{M}_d$. This density matrix encodes all the statistical information that can be obtained by applying repeated measurements on a sequence of n particles emitted by the source.

The amount of randomness in the process up to time n is quantified by the entropy

$$\mathsf{H}_n = \mathsf{H}\big(\omega_{\{0,1,\ldots,n\}}\big).$$

For stationary processes H_n satisfies sub-additivity $\mathsf{H}_{m+n} \leq \mathsf{H}_m + \mathsf{H}_n$. This implies the existence of the average entropy

$$\mathsf{h} := \lim_{n \to \infty} \frac{1}{n+1} \mathsf{H}_n. \tag{1}$$

The quantity h is also known as the dynamical entropy of the shift on the half-chain, it takes values in $[0, \log d]$. It is a relevant measure of the randomness of the source as, under suitable ergodicity conditions on $\boldsymbol{\omega}$, length-n messages can be reliably encoded in a subspace of dimension $\exp(n\mathsf{h})$ instead of d^n.

Using strong sub-additivity, $\mathsf{H}_{\ell+m+n} + \mathsf{H}_m \leq \mathsf{H}_{\ell+m} + \mathsf{H}_{m+n}$, more can be proven: the local entropy $n \mapsto \mathsf{H}_n$ is increasing, while the entropy increment $n \mapsto (\mathsf{H}_n - \mathsf{H}_{n-1})$ is decreasing. Both properties fail for general non shift-invariant quantum states. As a consequence

$$\mathsf{h} = \lim_{n \to \infty} (\mathsf{H}_n - \mathsf{H}_{n-1}). \tag{2}$$

This means that h is not only the compression rate of long messages but also the asymptotic entropy production of the source. The importance of (2) is that it can be used as a starting point for computing h. This happens for some Markov-like constructions where a simple transfer matrix-like construction generates the n-steps marginal of $\boldsymbol{\omega}$ from the $(n - 1)$-th. E.g., Blackwell[1] describes a procedure for computing the entropy of hidden Markov processes and we shall show that an analogous procedure applies to free Fermionic processes.

There are several routes that lead to classical Markov processes, like extending two-party states or generating the process in terms of a stochastic matrix. We remind here a number of problems and results that arise in this context with quantum processes. We also present a general scheme for generating quantum processes in terms of a quantum operation on a d-level system. Several examples are considered, in particular a free Fermionic version.

2. Classical Markov Processes

The configuration space of a classical register with d states is just a finite set $\Omega = \{1, 2, \ldots, d\}$. The states are length-$d$ probability vectors

$$\boldsymbol{\mu} = \{\mu(1), \mu(2), \ldots, \mu(d)\}, \quad \mu(\epsilon) \geq 0, \quad \sum_\epsilon \mu(\epsilon) = 1.$$

The state space is a simplex and the extreme points are the Dirac measures that assign a probability 1 to a configuration. The Shannon entropy

$$\mathsf{H}(\boldsymbol{\mu}) := -\sum_\epsilon \mu(\epsilon) \log \mu(\epsilon)$$

quantifies the randomness in the state. It is easily seen that H is a concave function on the state space:

$$H(\lambda_1 \boldsymbol{\mu}_1 + \lambda_2 \boldsymbol{\mu}_2) \geq \lambda_1 H(\boldsymbol{\mu}_1) + \lambda_2 H(\boldsymbol{\mu}_2), \quad \lambda_i \geq 0, \quad \lambda_1 + \lambda_2 = 1.$$

Restricting a state $\boldsymbol{\mu}_{12}$ on a composite system $\Omega_{12} = \Omega_1 \times \Omega_2$ to the subsystem Ω_1 returns the first marginal of $\boldsymbol{\mu}_{12}$

$$\boldsymbol{\mu}_1(\epsilon_1) := \sum_{\epsilon_2} \boldsymbol{\mu}_{12}(\epsilon_1, \epsilon_2).$$

The Shannon entropy behaves well with respect to restrictions:

- monotonicity: $H(\boldsymbol{\mu}_1) \leq H(\boldsymbol{\mu}_{12})$,
- sub-additivity: $H(\boldsymbol{\mu}_{12}) \leq H(\boldsymbol{\mu}_1) + H(\boldsymbol{\mu}_2)$, and
- strong sub-additivity: $H(\boldsymbol{\mu}_{123}) + H(\boldsymbol{\mu}_2) \leq H(\boldsymbol{\mu}_{12}) + H(\boldsymbol{\mu}_{23})$.

We can now consider the following state extension problem. Suppose that we are given two probability vectors $\boldsymbol{\mu}_{12}$ and $\boldsymbol{\nu}_{23}$ that agree on the middle system: $\boldsymbol{\mu}_2 = \boldsymbol{\nu}_2$. Can we find a joint extension for $\boldsymbol{\mu}_{12}$ and $\boldsymbol{\nu}_{23}$? More explicitly: can we find a state $\boldsymbol{\xi}_{123}$ on Ω_{123} that restricts to $\boldsymbol{\mu}_{12}$ on Ω_{12} and to $\boldsymbol{\nu}_{23}$ on Ω_{23}? This is indeed possible and clearly the set of joint extensions $\boldsymbol{\xi}_{123}$ is convex and compact. We can therefore refine the question and ask for a joint extension of maximal entropy. A straightforward computation yields the answer:

$$\boldsymbol{\rho}_{123}(\epsilon_1, \epsilon_2, \epsilon_3) := \frac{\boldsymbol{\mu}_{12}(\epsilon_1, \epsilon_2)\, \boldsymbol{\nu}_{23}(\epsilon_2, \epsilon_3)}{\boldsymbol{\mu}_2(\epsilon_2)} = \frac{\boldsymbol{\mu}_{12}(\epsilon_1, \epsilon_2)\, \boldsymbol{\nu}_{23}(\epsilon_2, \epsilon_3)}{\boldsymbol{\nu}_2(\epsilon_2)}. \tag{3}$$

Actually, $\boldsymbol{\rho}_{123}$ saturates the strong sub-additivity inequality:

$$H(\boldsymbol{\rho}_{123}) + H(\boldsymbol{\rho}_2) = H(\boldsymbol{\rho}_{12}) + H(\boldsymbol{\rho}_{23}).$$

Unsurprisingly, there is a direct connection with thermal equilibrium states. If we introduce Hamiltonians

$$\boldsymbol{\mu}_{12} = \mathrm{e}^{-h_{12}}, \quad \boldsymbol{\nu}_{23} = \mathrm{e}^{-h_{23}}, \quad \text{and } \boldsymbol{\mu}_2 = \boldsymbol{\nu}_2 = \mathrm{e}^{-h_2},$$

then

$$\boldsymbol{\rho}_{123} = \mathrm{e}^{-(h_{12}+h_{23}-h_2)}.$$

Let us start with a two-party probability vector $\boldsymbol{\mu}$ that is shift-invariant:

$$\sum_{\epsilon_2} \boldsymbol{\mu}(\epsilon, \epsilon_2) = \sum_{\epsilon_1} \boldsymbol{\mu}(\epsilon_1, \epsilon) \quad \text{for all } \epsilon. \tag{4}$$

We can repeatedly apply the Markov extension procedure (3) to get a stationary process

$$\omega(\epsilon_0, \epsilon_1, \dots, \epsilon_n) = \frac{\mu(\epsilon_0, \epsilon_1)\, \mu(\epsilon_1, \epsilon_2) \cdots \mu(\epsilon_{n-1}, \epsilon_n)}{\mu(\epsilon_1)\, \mu(\epsilon_2) \cdots \mu(\epsilon_{n-1})}. \tag{5}$$

Another procedure is to start with a $d \times d$ stochastic matrix T. The entry $T_{\epsilon_1 \epsilon_2}$ is the probability for jumping from state ϵ_1 to ϵ_2, therefore

$$T_{\epsilon_1 \epsilon_2} \geq 0 \text{ and } \sum_{\epsilon_2} T_{\epsilon_1 \epsilon_2} = 1. \tag{6}$$

The invariant state μ is a row vector determined by $\mu T = \mu$. The Markov process is now obtained by putting

$$\omega(\epsilon_0, \epsilon_1, \dots, \epsilon_n) = \mu(\epsilon_0)\, T_{\epsilon_0 \epsilon_1} \cdots T_{\epsilon_{n-1} \epsilon_n}. \tag{7}$$

Both constructions (5) and (7) agree if we put

$$T_{\epsilon_1 \epsilon_2} = \frac{\mu(\epsilon_1, \epsilon_2)}{\mu(\epsilon_1)}.$$

The rows of a stochastic matrix T are probability vectors. The minimal output entropy of T is simply

$$\mathsf{H}^{\min}(T) = \text{smallest entropy of rows of } T$$

while the entropy of the process is a smooth version of this quantity

$$\mathsf{h} = \mu\text{-average of entropies of rows of } T.$$

3. Extending Quantum States

When turning to quantum state extension the situation gets more complicated. Quantum states allow for more freedom, as they exhibit correlations that are not present in classical systems, but this imposes at the same time more stringent positivity conditions.

States on a full matrix algebra $\mathcal{M}_d$ can be identified with density matrices: non-negative matrices with trace one. The convex set of density matrices is very unlike a simplex. A density matrix that is not an extreme point of the state space, i.e., that is not a one-dimensional projector, allows many decompositions in extreme states. In contrast with classical systems such a state can therefore not be seen as a well-defined ensemble op pure states. We need $d^2 - 1$ real parameters to describe the state space of $\mathcal{M}_d$ while $2d - 2$ parameters suffice to label the pure states. This means that the boundary of the state space contains many flat subsets. Nevertheless the pure states form a very nice smooth manifold. The case of a single qubit is exceptional: its state space is affinely isomorphic to the Bloch ball by the standard parametrization

$$\rho = \tfrac{1}{2}(\mathbb{1} + \mathbf{x} \cdot \boldsymbol{\sigma}), \quad \mathbf{x} \in \mathbb{R}^3, \ \|\mathbf{x}\| \leq 1. \tag{8}$$

In this case, every point of the boundary is also an extreme point. For higher d, a smooth parametrization of the pure states does not define a boundary of a convex set.

For a composite system, restricting to a sub-system amounts to taking partial traces over remaining parties

$$\rho_1 := \mathrm{Tr}_2 \, \rho_{12}.$$

The entropy of a state with density matrix ρ is given by the von Neumann entropy

$$\mathsf{H}(\rho) = -\,\mathrm{Tr}\,\rho \log \rho.$$

However, already the most basic property of the Shannon entropy, monotonicity, does not carry over. Consider for example the maximally entangled two-qubit state $|\Phi^+\rangle\langle\Phi^+|$ with $|\Phi^+\rangle = (|00\rangle + |11\rangle)/\sqrt{2}$. Its entropy $\mathsf{H}(|\Phi^+\rangle\langle\Phi^+|)$ is zero as it is a pure state, while its restriction ρ_1 is the maximally mixed state, which has maximal entropy, so clearly $\mathsf{H}(\rho_1) \not\leq H(\rho_{12})$.

An important property that holds both for classical and quantum systems is that if the marginal ρ_1 of a bipartite state is pure then $\rho_{12} = \rho_1 \otimes \rho_2$. This is an important ingredient of the theory: it namely allows to isolate a system from the rest of the universe. At the same time it is also a severe constraint on quantum systems because there are plenty of pure states of a composite system. In particular the restriction of an entangled pure state can never be pure and we can therefore not separate a party of an entangled system from the outside world, which is more or less what goes wrong with the locality assumption in the EPR paradox.

Factorisation of extensions of pure states has also a bearing on joint extensions of states as considered in the previous section.[2] Indeed, suppose that ρ_{12} and ρ_{23} are pure and agree on the middle system, which is easily feasible, then a joint extension ρ_{123} can only exist for ρ_{12} and ρ_{23} pure product states. Therefore a generic pure two-party state with inner shift-invariance as in (4) cannot be extended.

Suppose that density matrices ρ_{12} and σ_{23} agree on the middle system and can be jointly extended. The set of extensions is still convex and compact and so we can still look for the maximal entropy extension. Finding this state is hard, however, because generally

$$\big[\rho_{12} \otimes \mathbb{1}_3 \,,\, \mathbb{1}_1 \otimes \sigma_{23}\big] \neq 0$$

or, equivalently, if ρ_{12} and σ_{23} are equilibrium states corresponding to Hamiltonians h_{12} and h_{23}

$$\mathrm{Tr}_3 \exp\big(h_{12} + h_{23}\big) \not\approx \exp\big(h_{12} + h_2\big).$$

Moreover, the maximal entropy extension will not saturate the strong sub-additivity.

Actually, a nice characterisation of equality in strong sub-additivity for a state ρ_{123} on a space $\mathcal{H}_1 \otimes \mathcal{H}_2 \otimes \mathcal{H}_3$ in terms of decompositions of the middle space has been obtained in the paper by Hayden et al.[3]. The necessary and sufficient condition is that the middle Hilbert space $\mathcal{H}_2$ decomposes as

$$\mathcal{H}_2 = \bigoplus_\alpha \mathcal{H}_{\mathrm{left}}^\alpha \otimes \mathcal{H}_{\mathrm{right}}^\alpha \ \text{ and } \ \rho_{123} = \bigoplus_\alpha \lambda_\alpha \, \rho_{12}^\alpha \otimes \rho_{23}^\alpha$$

with $\{\lambda_\alpha\}$ convex weights.

A Qubit Example with SU(2)-symmetry

An example of the limitations imposed on quantum state extensions can be worked out for qubits with a SU(2)-symmetry. In order to impose SU(2)-symmetry on single qubit observables we use the adjoint representation of SU(2)

$$\mathrm{Ad}(U) : A \mapsto U A U^*, \ \ U \in \mathrm{SU}(2), \ \ A \in \mathcal{M}_2.$$

This is a reducible representation that decomposes into a spin 0 and a spin 1 irrep:

$$\mathcal{M}_2 = \mathbb{C}\mathbb{1} \oplus \mathbb{C}\boldsymbol{\sigma}.$$

The only SU(2)-invariant state on $\mathcal{M}_2$ is the uniform state

$$\rho = \tfrac{1}{2}\,\mathbb{1}.$$

For 2 qubits $\mathrm{Ad}(U \otimes U)$ decomposes into 2 spin 0, 3 spin 1 and 1 spin 2 irrep. There exists now a one-parameter family of SU(2)-invariant states

$$\rho = \tfrac{1}{3}\,(1 - \lambda)(\mathbb{1} - \mathrm{p}) + \lambda\,\mathrm{p}, \ \ 0 \leq \lambda \leq 1.$$

Here p is the projector on the singlet vector $\frac{1}{\sqrt{2}}\,(|10\rangle - |01\rangle)$ in $\mathbb{C}^2 \otimes \mathbb{C}^2$. This projector commutes with every unitary of the form $U \otimes U$ and every two-qubit observable that is SU(2)-invariant is a linear combination of $\mathbb{p}$ and $\mathbb{1}$. Clearly, SU(2)-invariant two-qubit states satisfy

$$0 \leq \langle \mathbb{p} \rangle = \lambda \leq 1.$$

The two-qubit state ρ is separable for $0 \leq \lambda \leq \frac{1}{2}$ and entangled for $\frac{1}{2} < \lambda \leq 1$. Hence the expectation value of this projector for a certain process tells us how much bipartite entanglement between two neighbouring spins is attainable.

For 3 qubits the SU(2)-invariant states can still easily be determined but things become more complicated with increasing number of parties. Let

$$\mathbb{p}_1 = \mathbb{p} \otimes \mathbb{1} \text{ and } \mathbb{p}_2 = \mathbb{1} \otimes \mathbb{p} \text{ and put}$$
$$\mathbb{q} = \tfrac{4}{3} \left(\mathbb{p}_1 + \mathbb{p}_2 - \mathbb{p}_1 \mathbb{p}_2 - \mathbb{p}_2 \mathbb{p}_1 \right).$$

The algebra of three-qubit observables that are SU(2)-invariant is not Abelian. It can be decomposed into a direct sum of $\mathbb{C}$ and $\mathcal{M}_2$ where $\mathbb{C}$ is identified with $\mathbb{C}\mathbb{q}$ and $\mathcal{M}_2$ with the algebra generated by $\mathbb{p}_1$ and $\mathbb{p}_2$, not including $\mathbb{1}$. An SU(2)-invariant three-qubit state is of the form

$$\rho = \tfrac{1}{4}(1 - \lambda)(\mathbb{1} - \mathbb{q}) + \lambda \big(a\,\mathbb{p}_1 + b\,\mathbb{p}_2 + c\,\mathbb{p}_1\mathbb{p}_2 + \bar{c}\,\mathbb{p}_2\mathbb{p}_1 \big)$$

with

$$0 \leq \lambda \leq 1, \quad a, b \in \mathbb{R}, \quad c \in \mathbb{C}, \quad 2a + 2b + \Re\mathfrak{e}(c) = 1, \quad \text{and} \quad |c|^2 \leq 4ab.$$

If we look for a SU(2)-invariant three-party state with partial shift-invariance, then we find the following constraint on the expectation of $\mathbb{p}$

$$0 \leq \langle \mathbb{p}_1 \rangle = \langle \mathbb{p}_2 \rangle \leq 3/4. \tag{9}$$

SU(2) and shift-invariant states on more parties will satisfy stronger upper bounds on the expectations of $\mathbb{p}$, see (9). Ultimately, if we look for a shift-invariant extension on the full half-chain then, using the Bethe Ansatz[4], one can show that

$$0 \leq \langle \mathbb{p} \rangle \leq \log 2 \approx 0.69. \tag{10}$$

We may look for the largest expectation value of $\mathbb{p}$ that can be obtained within classes of shift-invariant states that can easily be handled. Consider as a first example point-wise limits of shift-invariant product states. Such states are actually invariant under arbitrary finite permutations of sites on the half-chain and are usually called exchangeable. Using the Bloch parametrization (8) and

$$\mathbb{p} = \tfrac{1}{4}(\mathbb{1} - \boldsymbol{\sigma}_1 \cdot \boldsymbol{\sigma}_2) \text{ with } \boldsymbol{\sigma}_1 = \boldsymbol{\sigma} \otimes \mathbb{1} \text{ and } \boldsymbol{\sigma}_2 = \mathbb{1} \otimes \boldsymbol{\sigma}$$

we have to maximize

$$\mathbf{x} \in \mathbb{R}^3 \mapsto \tfrac{1}{16} \, \mathrm{Tr}\big[(\mathbb{1} + \mathbf{x} \cdot \boldsymbol{\sigma}_1)(\mathbb{1} + \mathbf{x} \cdot \boldsymbol{\sigma}_2)(\mathbb{1} - \boldsymbol{\sigma}_1 \cdot \boldsymbol{\sigma}_2)\big]$$

subject to the constraint $\|\mathbf{x}\| \leq 1$. It is easily seen that the maximum is reached for $\mathbf{x} = 0$ for which value $\langle \mathbb{p} \rangle = \tfrac{1}{4}$. Hence

$$\langle \mathbb{p} \rangle \leq \tfrac{1}{4} \text{ for exchangeable states.} \tag{11}$$

The largest expectation for $\mathbb{p}$ that can be reached within the class of product states is

$$\max\Big(\tfrac{1}{16} \, \mathrm{Tr}\big[(\mathbb{1} + \mathbf{x}_1 \cdot \boldsymbol{\sigma}_1)(\mathbb{1} + \mathbf{x}_2 \cdot \boldsymbol{\sigma}_2)(\mathbb{1} - \boldsymbol{\sigma}_1 \cdot \boldsymbol{\sigma}_2)\big]\Big)$$

subject to the constraint $\|\mathbf{x}_1\|, \|\mathbf{x}_2\| \leq 1$. The maximum $\tfrac{1}{2}$ is attained for $\mathbf{x}_1 = -\mathbf{x}_2 = \mathbf{x}$ where $\mathbf{x} \in \mathbb{R}^3$ is an arbitrary vector of length 1. Therefore

$$\langle \mathbb{p} \rangle \leq \tfrac{1}{2} \text{ for separable states.}$$

Moreover, this maximum is attained for shift-invariant separable states that are equal weight mixtures of period-2 product states

$$\tfrac{1}{2} \, |e_0\rangle\langle e_0| \otimes |e_1\rangle\langle e_1| \otimes |e_0\rangle\langle e_0| \otimes \cdots + \tfrac{1}{2} \, |e_1\rangle\langle e_1| \otimes |e_0\rangle\langle e_0| \otimes |e_1\rangle\langle e_1| \otimes \cdots , \tag{12}$$

where $\{e_0, e_1\}$ is any orthonormal basis in $\mathbb{C}^2$. Hence

$$\langle \mathbb{p} \rangle \leq \tfrac{1}{2} \text{ for shift-invariant separable states}$$

is an optimal upper bound. States of the form (12) are extreme shift-invariant states which allow a convex decomposition in clustering period-2 states. This is called Néel order of period 2. The value $\tfrac{1}{2}$ for shift-invariant separable states is still not close to the maximum value of $\log 2$. One can get closer by constructing more general quantum processes.

4. Constructing Processes

We now turn to the construction of classical and quantum processes using as initial data a unity preserving CP map $\Gamma : \mathcal{M}_d \to \mathcal{M}_d$ with invariant state ρ. In the classical case this reduces to a stochastic matrix T with invariant measure μ. The construction is based on finitely correlated states[5], also called matrix product states. These are processes for which the correlations across any link can be modelled by a finite dimensional vector space. These states are more general than the ones we have considered until now and are easily constructible in the thermodynamic limit, unlike the Bethe Ansatz states. The finitely correlated states where e.g. used in the lectures by J. Eisert under the form of tensor networks. Actually pure states have been considered there as these lectures were focusing on ground states. We present here a different version that is adapted to mixed states.

The starting point is a unity preserving completely positive (UPCP) map

$$\Lambda : \mathcal{M}_d \otimes \mathcal{M}_d \to \mathcal{M}_d$$

that is compatible with the given Γ in the following sense

$$\Lambda(A \otimes \mathbb{1}) = \Lambda(\mathbb{1} \otimes A) = \Gamma(A), \quad A \in \mathcal{M}_d. \tag{13}$$

A process ω is now generated by repeatedly contracting the local observables on the half-chain. Consider a sequence of UPCP maps

$$\Lambda^{(0)} := \Lambda : \mathcal{M}_d \otimes \mathcal{M}_d \to \mathcal{M}_d$$

$$\Lambda^{(1)} := \Lambda \circ (\Lambda \otimes \mathrm{id}) : \mathcal{M}_d \otimes (\mathcal{M}_d \otimes \mathcal{M}_d) \to \mathcal{M}_d$$

$$\vdots \tag{14}$$

$$\Lambda^{(n)} := \Lambda \circ (\Lambda^{(n-1)} \otimes \mathrm{id}) : \mathcal{M}_d \otimes \underbrace{(\mathcal{M}_d \otimes \cdots \otimes \mathcal{M}_d)}_{(n+1) \text{ times}} \to \mathcal{M}_d.$$

The expectation of a local observable $A_n \in \otimes_0^n \mathcal{M}_d$ is then computed as

$$\omega(A_n) := \mathrm{Tr}\Big\{\rho \Lambda^{(n)}(\mathbb{1} \otimes A_n)\Big\}. \tag{15}$$

To define a stationary process, (15) must satisfy a number of requirements. The definition should be consistent in the first place, namely $\omega(A_n \otimes \mathbb{1}) = \omega(A_n)$. This follows from the compatibility (13) and the invariance of ρ:

$$\omega(A_n \otimes \mathbb{1}) = \mathrm{Tr}\Big\{\rho \Lambda^{(n+1)}(\mathbb{1} \otimes A_n \otimes \mathbb{1})\Big\}$$

$$= \mathrm{Tr}\Big\{\rho \left(\Lambda \circ (\Lambda^{(n)} \otimes \mathrm{id})\right)(\mathbb{1} \otimes A_n \otimes \mathbb{1})\Big\}$$

$$= \mathrm{Tr}\Big\{\rho \Lambda\big(\Lambda^{(n)}(\mathbb{1} \otimes A_n) \otimes \mathbb{1}\big)\Big\}$$

$$= \mathrm{Tr}\Big\{\rho \Gamma\big(\Lambda^{(n)}(\mathbb{1} \otimes A_n)\big)\Big\}$$

$$= \mathrm{Tr}\Big\{\rho \Lambda^{(n)}(\mathbb{1} \otimes A_n)\Big\}$$

$$= \omega(A_n).$$

Next, we need positivity. This follows immediately from the complete positivity of Λ. The compatibility condition implies that Λ maps the identity on $\mathbb{C}^d \otimes \mathbb{C}^d$ to the identity on $\mathbb{C}^d$. This implies the normalization and stationarity of ω.

It is important to observe that compatibility (13) imposes a severe restriction on Γ. Not every UPCP transformation Γ of $\mathcal{M}_d$ admits a compatible extension. Moreover, the compatible extensions of Γ, whenever such extensions exist, form a compact and convex set and one may wonder about particular extensions. We now turn to some classes of examples.

4.1. Hidden Markov Processes

A classical observable, i.e., a $\mathbb{R}$-valued function f on configuration space $\Omega = \{1, 2, \ldots, d\}$ is naturally tabulated into a vector $\mathbf{f} = \big(f(1), f(2), \ldots, f(d)\big)^{\mathsf{T}} \in \mathbb{R}^d$ and identified with a diagonal matrix in $\mathcal{M}_d$ through the map

$$\mathrm{dia}(\mathbf{f}) = \sum_{\epsilon} f(\epsilon) \, |\epsilon\rangle\langle\epsilon|.$$

The relation between a (completely) positive transformation Γ of $\mathcal{M}_d$ and a stochastic $d \times d$ matrix is then

$$\Gamma\big(\mathrm{dia}(\mathbf{f})\big) = \mathrm{dia}(T\,\mathbf{f}).$$

This allows to rewrite the compatibility equation (13): a stochastic matrix $S : \mathbb{R}^d \otimes \mathbb{R}^d \to \mathbb{R}^d$ is compatible with a stochastic matrix $T : \mathbb{R}^d \to \mathbb{R}^d$ if

$$\sum_{\epsilon_2} S_{\varphi,(\epsilon,\epsilon_2)} = \sum_{\epsilon_1} S_{\varphi,(\epsilon_1,\epsilon)} = T_{\varphi,\epsilon}, \ \forall \ \varphi, \epsilon.$$

Let us introduce d square matrices of dimension d with non-negative entries

$$E(\epsilon)_{\varphi,\eta} = S_{\varphi,(\eta,\epsilon)}.$$

The process generated by S is then seen to be

$$\boldsymbol{\omega}(\epsilon_0, \epsilon_1, \ldots, \epsilon_n) = \langle \boldsymbol{\mu}, \, E(\epsilon_0)E(\epsilon_1) \cdots E(\epsilon_n)\mathbf{1}\rangle,$$

where $\mathbf{1} \in \mathbb{R}^d$ has all its entries equal to one and $\boldsymbol{\mu}$ is the invariant probability vector for T.

A stochastic matrix T always allows the extension

$$S_{\varphi,(\eta,\epsilon)} = \delta_{\eta,\epsilon}\, T_{\varphi,\epsilon}.$$

The corresponding process is the usual Markov process (7). More general extensions $\boldsymbol{\omega}$ are hidden Markov processes: there exists a larger configuration space Ω_1, a function $F : \Omega_1 \to \Omega$ and a Markov process $\boldsymbol{\omega}_1$ on Ω_1 such that

$$\boldsymbol{\omega}(\epsilon_0, \epsilon_1, \ldots, \epsilon_n) = \sum_{F(\varphi_j)=\epsilon_j} \boldsymbol{\omega}_1(\varphi_0, \varphi_1, \ldots, \varphi_n).$$

The entropy of hidden Markov processes can be computed using a method due to Blackwell[1,6]. The starting point is the asymptotic entropy production formula (2). The construction of the process, adding one point at a time, see (14) and (15), defines a dynamical system on the length-d probability vectors. The entropy of the process is then obtained as an average over entropies of probability vectors with respect to the invariant measure of the dynamical system. Numerical evidence suggests that the Markov extension has the smallest entropy amongst all.

4.2. *Qubits with SU(2)-invariance cont.*

In order to have manifest SU(2)-invariance of the process we impose SU(2)-covariance both on the CP transformation of $\mathcal{M}_2$ and on its compatible extensions from $\mathcal{M}_2 \otimes \mathcal{M}_2$ to $\mathcal{M}_2$. Let $\mathcal{G} \mapsto U_g$ be a unitary representation of a group $\mathcal{G}$ on a Hilbert space $\mathcal{H}$. The adjoint representation lifts it to a representation of $\mathcal{G}$ on the bounded linear transformations $\mathcal{B}(\mathcal{H})$ of $\mathcal{H}$:

$$\mathrm{Ad}(U_g)(A) = U_g A U_g^*, \quad g \in \mathcal{G}, \ A \in \mathcal{B}(\mathcal{H}).$$

Given two unitary representations $U^{(1)}$ and $U^{(2)}$ of $\mathcal{G}$ on $\mathcal{H}_1$ and $\mathcal{H}_2$ a map $\Gamma : \mathcal{B}(\mathcal{H}_1) \to \mathcal{B}(\mathcal{H}_2)$ is covariant if

$$\Gamma \circ \mathrm{Ad}(U^{(1)}) = \mathrm{Ad}(U^{(2)}) \circ \Gamma \tag{16}$$

The Choi-Jamiołkowski encoding of a linear map $\Gamma : \mathcal{M}_{d_1} \to \mathcal{M}_{d_2}$ is very convenient for handling complete positivity

$$\mathsf{C}(\Gamma) := \sum_{i,j} |i\rangle\langle j| \otimes \Gamma(|i\rangle\langle j|),$$

Γ is completely positive if and only if $\mathsf{C}(\Gamma)$ is positive semi-definite. The encoding depends on the chosen basis through the matrix units $|i\rangle\langle j|$ but only up to unitary equivalence as

$$\mathsf{C}(\Gamma \circ \mathrm{Ad}(U)) = \mathrm{Ad}(U^\mathsf{T} \otimes \mathbb{1}) \circ \mathsf{C}(\Gamma) \text{ and}$$
$$\mathsf{C}(\mathrm{Ad}(U) \circ \Gamma) = \mathrm{Ad}(\mathbb{1} \otimes U) \circ \mathsf{C}(\Gamma).$$

The covariance condition (16) for $\Gamma : \mathcal{M}_{d_1} \to \mathcal{M}_{d_2}$ translates for its Choi-Jamiołkowski encoding into

$$[\overline{U}_g^{(1)} \otimes U_g^{(2)}, \mathsf{C}(\Gamma)] = 0, \quad g \in \mathcal{G}.$$

Here $\overline{A}$ is the complex conjugate of the matrix A. For SU(2) there is an additional simplification because the conjugate of SU(2) is unitarily equivalent to SU(2).

It turns out that there is a one-parameter family of SU(2)-covariant UPCP transformations of $\mathcal{M}_2$

$$\Gamma(\boldsymbol{\sigma}) = \mu\,\boldsymbol{\sigma}, \quad -\tfrac{1}{3} \le \mu \le 1.$$

The SU(2)-covariant UPCP maps $\Lambda : \mathcal{M}_2 \otimes \mathcal{M}_2 \to \mathcal{M}_2$ compatible with Γ are parametrized by three real parameters

$$
\begin{aligned}
\Lambda(\boldsymbol{\sigma}_1 \cdot \boldsymbol{\sigma}_2) &= \alpha\,\mathbb{1}, \\
\Lambda(\boldsymbol{\sigma}_1) = \Lambda(\boldsymbol{\sigma}_2) = \Gamma(\boldsymbol{\sigma}) &= \mu\,\boldsymbol{\sigma}, \quad \text{and} \\
\Lambda(\boldsymbol{\sigma}_1 \times \boldsymbol{\sigma}_2) &= \eta\,\boldsymbol{\sigma}.
\end{aligned}
\tag{17}
$$

Complete positivity imposes constraints on α, μ, and η

$$|6\mu - \alpha| \le 3 \quad \text{and} \quad 3 - 2\alpha - \alpha^2 + 12\mu - 12\alpha\mu - 9\eta^2 \ge 0. \tag{18}$$

These conditions can be obtained by imposing positivity on the Choi matrix of Λ. The allowed region is a piece of a cone in $\mathbb{R}^3$. We then compute the expectation of $\mathbb{p}$

$$
\begin{aligned}
\langle \mathbb{p} \rangle &= \tfrac{1}{4} - \tfrac{1}{4}\langle \boldsymbol{\sigma}_1 \cdot \boldsymbol{\sigma}_2 \rangle = \tfrac{1}{4} - \tfrac{1}{8}\,\mathrm{Tr}\sum_\gamma \Lambda\big(\sigma^\gamma \otimes \Lambda(\sigma^\gamma \otimes \mathbb{1})\big) \\
&= \tfrac{1}{4} - \tfrac{1}{8}\,\mu\,\mathrm{Tr}\,\Lambda(\boldsymbol{\sigma}_1 \cdot \boldsymbol{\sigma}_2) = \tfrac{1}{4}(1 - \alpha\mu).
\end{aligned}
$$

The maximum in the allowed parameter region is attained for $\alpha = -\tfrac{3}{2}$ and $\mu = \tfrac{1}{4}$ and is independent of η. Therefore

$$\langle \mathbb{p} \rangle \le \tfrac{11}{32} \tag{19}$$

for $\langle\ \rangle$ a stationary and SU(2)-invariant process as in (14). This should be compared with (11).

In passing from exchangeable to shift-invariant separable states we actually allowed product states of period 2. This can also be applied to processes of the type (17). Considering $\boldsymbol{\sigma}_1 \cdot \boldsymbol{\sigma}_2$ as the contribution to the energy of two neighbouring spins, a minimal value of $\langle \boldsymbol{\sigma}_1 \cdot \boldsymbol{\sigma}_2 \rangle$ corresponds to a maximal value of $\langle \mathbb{p} \rangle$ and this is expected to happen for spins as anti-parallel as possible. Therefore the second requirement in (17) is inappropriate and we should consider general SU(2)-covariant maps $\Lambda : \mathcal{M}_2 \otimes \mathcal{M}_2 \to \mathcal{M}_2$. These are determined by four real parameters

$$
\begin{aligned}
\Lambda(\boldsymbol{\sigma}_1 \cdot \boldsymbol{\sigma}_2) &= \alpha\,\mathbb{1}, \\
\Lambda(\boldsymbol{\sigma}_1) = \mu\,\boldsymbol{\sigma}, \quad \Lambda(\boldsymbol{\sigma}_2) &= \nu\,\boldsymbol{\sigma}, \quad \text{and} \\
\Lambda(\boldsymbol{\sigma}_1 \times \boldsymbol{\sigma}_2) &= \eta\,\boldsymbol{\sigma}.
\end{aligned}
\tag{20}
$$

Complete positivity imposes the constraints

$$|3\mu + 3\nu - \alpha| \leq 3 \text{ and}$$
$$3 - 2\alpha - \alpha^2 + 6(1 - \alpha)(\mu + \nu) - 9(\mu - \nu)^2 - 9\eta^2 \geq 0. \tag{21}$$

We now introduce two SU(2)-covariant maps $\Lambda_i : \mathcal{M}_2 \otimes \mathcal{M}_2 \to \mathcal{M}_2$ as in (20) with defining parameters $(\alpha_i, \mu_i, \nu_i, \eta_i)$. The expectation of $\mathbb{p}$ in the equal weight average of these period-2 processes is given by

$$\langle \mathbb{p} \rangle = \tfrac{1}{4} - \tfrac{1}{8}\left(\alpha_2\mu_1 + \alpha_1\nu_2\right).$$

Maximizing this in the allowed parameter region yields

$$\langle \mathbb{p} \rangle = \tfrac{5}{8} = 0.625$$

which is within 10% of the optimal bound and well within the entangled shift-invariant states.

4.3. *Davies Maps*

An interesting and physically relevant class of channels are the Davies maps, they arise in the reduced dynamical description of a system with a discrete level structure weakly coupled to a thermal bath[7]. The level structure of the small system is preserved in the sense that such a map Γ is parametrized by a stochastic map T and a decoherence matrix D. The matrix T describes the stochastic evolution of the diagonal elements while D gives the damping of the off-diagonal terms. Assuming that the system Hamiltonian is diagonal in the canonical basis

$$\Gamma\big(\mathrm{dia}(\varphi)\big) = \mathrm{dia}(T\,\varphi) \text{ and } \Gamma(e_{ij}) = D_{ij}e_{ij}, \ i \neq j. \tag{22}$$

Here, $e_{ij} = |e_i\rangle\langle e_j|$. Moreover, T is detailed balance and D is real symmetric. Detailed balance means that T is Hermitian for the stationary measure $\boldsymbol{\mu}$ that is interpreted as the Gibbs state of the system

$$\boldsymbol{\mu}(f\,T(g)) = \boldsymbol{\mu}(T(f)\,g), \ f \text{ and } g \text{ real-valued.}$$

This condition is equivalent with micro-reversibility: the occupation rate of level i times the jump probability from i to j is equal to the occupation rate of j times the jump probability from j to i. Another equivalent condition is

$$T_{ij}T_{jk}T_{ki} = T_{ik}T_{kj}T_{ji} \text{ for all choices of } i, j, k.$$

Complete positivity additionally imposes that

$$\begin{pmatrix} T_{11} & D_{12} & \cdots & D_{1d} \\ D_{21} & T_{22} & \cdots & D_{2d} \\ \vdots & \vdots & \ddots & \vdots \\ D_{d1} & D_{d2} & \cdots & T_{dd} \end{pmatrix} \quad \text{be positive semi-definite.} \qquad (23)$$

The action of Γ is quite clear: decay of the off-diagonal elements and birth and death process for the diagonals. The relations between the different rates are encoded in the positivity condition (23). E.g., one can readily check that the decay rate of the off-diagonals cannot be less than half the rate of convergence to equilibrium for the diagonal process.

For Davies maps one could expect the standard basis vectors to be minimizers of output entropy but this is not generally true. It has been shown[8] that already for a single qubit a true superposition of ground and excited state is the minimizer in a regime where the map is close to the identity map and so truly quantum. High powers of a Davies map converge to the projector on the equilibrium state which is entanglement breaking. In this regime the minimizer for output entropy is the state corresponding to the row of minimal entropy in T.

The construction of a process as in (14) requires a Davies map rather closer to the projector on the equilibrium state than to the identity. For a single qubit

$$T = \begin{pmatrix} 1-a & a \\ b & 1-b \end{pmatrix}, \quad \text{with } 0 \le b \le a \le 1$$

and with d the damping factor of the off-diagonal element one checks that it generates a process if

$$d^2 \le \tfrac{1}{2}(1-a)(1-b).$$

4.4. *Free Fermionic Processes*

For both Bosons and Fermions there exists a notion of Gaussian states and maps[9,10] that are considerably simpler to handle than general ones. The names free, quasi-free, quadratic, linear, and determinantal are also used. Moreover, these states and maps are good approximations whenever the statistics dominates over the interactions. A considerable benefit is also the scaling behaviour: the dimension of the free objects grows linearly in the number of particles instead of exponentially. We shall here only describe the defining free objects without connecting them to the true observables, states, and maps of a many particle system. This yields a kind of meta-description. Moreover, we restrict ourselves to Fermions.

The observables of Fermions with mode space $\mathcal{H}$ are the trace-class operators $\mathcal{I}_1(\mathcal{H})$ on $\mathcal{H}$. Apart from their linear structure commutators are also useful. The mode space $\mathcal{H}_{12}$ of a bipartite system is just the direct sum of the mode spaces $\mathcal{H}_1$ and $\mathcal{H}_2$ of the corresponding subsystems and observables of system 1 are extended by putting

$$A \in \mathcal{I}_1(\mathcal{H}_1) \mapsto A \oplus 0 \in \mathcal{I}_1(\mathcal{H}_{12}).$$

The symbols play the role of states, they are operators Q on $\mathcal{H}$ such that $0 \leq Q \leq \mathbb{1}$. The expectation of an observable A is just $\operatorname{Tr} Q A$. Mixtures of symbols are constrained by the following requirement: let $0 < \lambda < 1$ then the mixture of $Q_1 \neq Q_2$ with weights λ and $1 - \lambda$ can be formed if and only if $Q_1 - Q_2$ is of rank 1 and it yields the symbol $\lambda Q_1 + (1 - \lambda)Q_2$. It then follows that a symbol is pure if and only if it is an orthogonal projector, possibly 0. Given a composite system $\mathcal{H}_{12} = \mathcal{H}_1 \oplus \mathcal{H}_2$ and two symbols Q_1 and Q_2, the product state has symbol $Q_1 \oplus Q_2$. In this context, a separable state is just block diagonal in the mode space decomposition. We shall also need the von Neumann entropy of a symbol on a finite dimensional space

$$\mathsf{H}(Q) := -\operatorname{Tr}\big[Q \log Q + (\mathbb{1} - Q) \log(\mathbb{1} - Q)\big]. \tag{24}$$

In particular, $\mathsf{H}(Q) = 0$ if and only if Q is pure.

A trace-preserving completely positive map from $\mathcal{H}_1 \to \mathcal{H}_2$ is a couple (A, B) of linear maps where

$$A : \mathcal{H}_1 \to \mathcal{H}_2, \ B : \mathcal{H}_1 \to \mathcal{H}_1, \ \text{and} \ 0 \leq B \leq \mathbb{1} - A^*A. \tag{25}$$

The action on a symbol is given by

$$Q \mapsto Q' = A^*Q A + B.$$

Observe that such maps are compatible with the notion of convex mixture of above because there is only a single Kraus-like operator appearing in (25). Composition of free CP maps is given by a semi-direct product

$$(A, B) \circ (A', B') = (A A', B' + (A')^*B A').$$

We now mimic within the Fermionic free context the construction of a process starting from a CP transformation (A, B) of $\mathbb{C}^d$. Such a map can be extended to a compatible map (C, D) from $\mathbb{C}^d \oplus \mathbb{C}^d \to \mathbb{C}^d$ if and only if $A^*A \leq \min(\{\frac{1}{2}\mathbb{1}, \mathbb{1} - B\})$ and the extensions are labelled by an $X : \mathbb{C}^d \to \mathbb{C}^d$

$$C = (A \ A) \quad \text{and} \quad D = \begin{pmatrix} B & X \\ X^* & B \end{pmatrix}.$$

The symbol Q invariant under (A, B) is the solution of

$$Q = A^*Q\,A + B.$$

It is not hard to see that the outcome of the construction is a symbol Q_∞ on $\oplus_0^\infty \mathbb{C}^d$ which is a block Toeplitz matrix: the $d \times d$ matrix entries in Q_∞ are constant along parallels to the main diagonal. Explicitly

$$(Q_\infty)_{ii} = Q \quad \text{and} \quad (Q_\infty)_{i\,i+n} = (A^*)^n(Q - B + X). \tag{26}$$

The entropy can now be computed in two different ways, we can either compute the limiting average entropy as in (1) or the asymptotic entropy production as in (2). The first method relies on an extension of Szegö's theorem to block Toeplitz matrices. For the second we need either a much finer control on the spectra of principal sub-matrices of a block Toeplitz matrix, which appears to be hard, or we have to exploit the smoothness of the entropy function. We follow this last approach.

Let $T : [-\pi, \pi[\to \mathcal{M}_d$ be an $\mathcal{L}^\infty$-function taking values in the Hermitian $d \times d$ matrices and put

$$\hat{T} = \begin{pmatrix} \hat{T}(0) & \hat{T}(1) & \hat{T}(2) & \cdots \\ \hat{T}(-1) & \hat{T}(0) & \hat{T}(1) & \cdots \\ \hat{T}(-2) & \hat{T}(-1) & \hat{T}(0) & \cdots \\ \vdots & \vdots & \vdots & \ddots \end{pmatrix}$$

where $\hat{T}$ are the Fourier coefficients of T

$$\hat{T}(k) := \tfrac{1}{2\pi} \int_{-\pi}^{\pi} d\theta\, T(\theta)\, e^{-2\pi i k\theta}.$$

An extension of the classical Szegö theorem[11] reads

Theorem 4.1 (Szegö). *For any continuous complex function f on $\mathbb{R}$*

$$\lim_{n\to\infty} \frac{1}{n}\, \mathrm{Tr}\, f(P_n\hat{T}\,P_n) = \frac{1}{2\pi} \int_{-\pi}^{\pi} d\theta\, \mathrm{Tr}\, f(T(\theta)) \tag{27}$$

where P_n projects onto the first n terms in $\oplus_0^\infty \mathbb{C}^d$.

The theorem gives us information on the main asymptotic behaviour of the eigenvalues of principal sub-matrices of $\hat{T}$. Consider e.g., the case $d = 1$, then Szegö's theorem can be rewritten as

$$\lim_{n\to\infty} \frac{1}{n}\, \mathrm{Tr}\, f(P_n\hat{T}\,P_n) = \int_{\mathbb{R}} \mu(dx)\, f(x)$$

where

$$\mu(] - \infty, x]) = \frac{1}{2\pi} \int_{T(\theta) \leq x} d\theta.$$

If we order the eigenvalue list $\{\lambda_{n,j}\}$ of $P_n \hat{T} P_n$, then $\{\lambda_{n,j}\}$ interlaces $\{\lambda_{n+1,j}\}$ and

$$\text{w}^*\text{-} \lim \frac{1}{n} \sum_{j=1}^{n} \delta_{\lambda_{n,j}} = \mu.$$

A fine asymptotic control on the eigenvalues could be used to obtain the average (27) as an asymptotic growth

$$\lim_{n \to \infty} \left\{ \text{Tr} \, f(P_n \hat{T} \, P_n) - \text{Tr} \, f(P_{n-1} \hat{T} \, P_{n-1}) \right\}.$$

Numerical evidence, however, shows that the behaviour of eigenvalue spacings can become erratic when T oscillates.

In fig. 1 the function $T(\theta) = \frac{1}{2} + \frac{1}{5} \cos(\theta) + \frac{1}{3} \sin(2\theta)$ is plotted. In fig. 2 the eigenvalue lists of the first 50 principal sub-matrices are shown together with a plot of the eigenvalues of the 100×100 sub-matrix. This last plot approximates well the reordered function T.

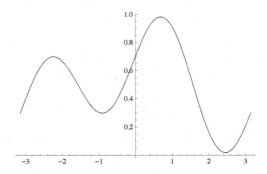

Figure 1.   The function $T(\theta) = \frac{1}{2} + \frac{1}{5} \cos(\theta) + \frac{1}{3} \sin(2\theta)$.

Strong sub-additivity of the entropy guarantees the existence of the asymptotic entropy production but a much more general result can be proven by extending Szegö's theorem to reasonably smooth functions. This result can then be applied to the computation of the entropy of our processes.

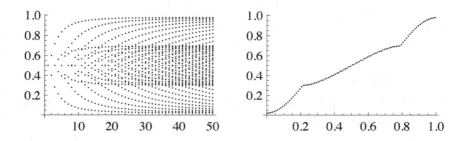

Figure 2.   Eigenvalues of principal Toeplitz sub-matrices.

Theorem 4.2. *For a block Toeplitz matrix T and an absolutely continuous complex function f on $\mathbb{R}$*

$$\lim_{n \to \infty} \left(\mathrm{Tr}\, f(P_{n+1} \hat{T} P_{n+1}) - \mathrm{Tr}\, f(P_n \hat{T} P_n) \right) = \frac{1}{2\pi} \int_{-\pi}^{\pi} d\theta\, \mathrm{Tr}\, f(T(\theta)).$$

As a corollary we get

$$\mathsf{h} = \tfrac{1}{2\pi} \int_{-\pi}^{\pi} d\theta\, \mathsf{H}(Q_\infty(\theta))$$

with

$$Q_\infty = Q + \frac{A^* e^{i\theta}}{\mathbb{1} - A^* e^{i\theta}} \left(Q - B + X \right) + \mathrm{h.c.}$$

and H as in (24).

5. Conclusion

The construction of quantum processes is a lot less straightforward than for their classical counterparts. The intricate nature of quantum correlations complicates even the seemingly simple task of finding extensions of overlapping states.

Processes that can nevertheless be easily constructed, like exchangeable or separable states, are not general enough to study interesting quantum behaviour. On the other hand, more general constructions like the Bethe Ansatz become difficult to handle as the size of the process increases.

The processes we have studied here, the finitely correlated states, lie somewhere in between the previous two classes. By construction, they are

well-behaved as the length of the process grows. We have also seen by studying some concrete examples that such states do in fact exhibit interesting
quantum characteristics.

Acknowledgements

This work is partially funded by the Belgian Interuniversity Attraction
Poles Programme P6/02.

Bibliography

1. D. Blackwell: The entropy of functions of finite state Markov chains, Trans-
   actions of the first Prague Conference on Information Theory, Statistical De-
   cision Functions, Random Processes, Publishing House of the Czechoslovak
   Academy of Sciences (1957)
2. R.F. Werner: Remarks on a quantum state extension problem, Lett. Math.
   Phys. **19**, 319–26 (1990)
3. P. Hayden, R. Jozsa, D. Petz, and A. Winter: Structure of states which satisfy strong subadditivity of quantum entropy with equality, Commun. Math.
   Phys. **246**, 359–74 (2004)
4. H. Bethe: Zur Theorie der Metalle I. Eigenwerte und Eigenfunktionen Atom-
   kete, Z. Phys. **71**, 205–226 (1931)
5. M. Fannes, B. Nachtergaele, and R.F. Werner: Finitely correlated states on
   quantum spin chains, Commun. Math. Phys. **144**, 443–90 (1992)
6. M. Fannes, B. Nachtergaele, and L. Slegers: Functions of Markov processes
   and algebraic measures, Rev. Math. Phys. **4**, 39–64 (1992)
7. R. Alicki and K. Lendi: Quantum Dynamical Semigroups and Applications,
   Springer LNP **717** (2007)
8. W. Roga, M. Fannes, and K. Życzkowski: Davies maps for qubits and qutrits,
   arXiv:0911.5607[quant-ph].
9. R. Alicki and M. Fannes: Quantum Dynamical Systems, Oxford University
   Press, Oxford (2001)
10. O. Bratteli and D. W. Robinson: Operator Algebras and Quantum Statistical
    Mechanics 2: Equilibrium States. Models in Quantum Statistical Mechanics,
    Springer, Berlin (1997)
11. M. Miranda and P. Tilli: Asymptotic spectra of Hermitian block Toeplitz
    matrices and preconditioning results, SIAM J. Matrix Anal. Appl. **21**, 867–
    881 (2000)

Pure state entanglement in terms of nilpotent variables: η-toolbox

Andrzej Frydryszak*

*Institute of Theoretical Physics, University of Wrocław,
Wrocław, Poland*
* E-mail: amfry@ift.uni.wroc.pl

Entanglement is one of the most important resources in developing quantum computer technology. It might be surprising that it can be effectively described in spaces of functions of nilpotent variables η. For such functions in a natural way arise collections of invariants allowing to construct entanglement measures. Moreover various examples of nontrivial entangled states obtained using quantum physical reasoning turn out to be naturally related to the elementary functions of η-variables.

Keywords: nilpotent quantum mechanics; classical invariant theory; entanglement; entanglement monotones.

1. Introduction

Since its very beginning quantum mechanics is one of the most important areas of scientific activity, not only physical, but also mathematical, influencing the development of various mathematical formalisms, frequently of importance for mathematics itself, such as: functional analysis, theory of distributions, operator theory, C^*-algebras, complex projective spaces, harmonic analysis on the Heisenberg group and induced representations, theory of deformations. Its popularity in the last two decades is coupled in great extend to the immense advancement of the experimental techniques. Among others, it is related to development of quantum information theory and a new resource we have found in Nature - the quantum entanglement. It turned out, that mathematical questions arising in this context are essentially nontrivial, some of them were posed long time ago. All these issues locate quantum mechanics again at the frontier of theoretical physics, in a position, once occupied, then overtaken for decades by fundamental interaction theories - (super)string theory, quantum gravity, supergravity, quantum field theory conventional and supersymmetric, QED and QCD. Is needless to say that the experimental devices needed for penetrating the Nature in this regime are of absolutely noncomparable size to that of the LHC, but the joy of discovering new properties of the Nature, is comparable.

Moreover, mathematical tools needed to study above mentioned fundamental theories, are very sophisticated and abstract, in view of that, problems met in the contemporary non-relativistic quantum mechanics might seem to be "easy". In fact they are nontrivial, fundamental and some of them are open by now. The description of quantum entanglement is linked to a collection of such problems rooted in geometry, linear algebra, C^*-algebras and functional analysis. Some of long standing questions have very simple formulation, what might be misleading, beacause they are deep fundamental problems, frequently not answered by now.

In the present exposition we wish to take a view onto the entanglement problem of multipartite systems in terms of functions of nilpotent variables. Such approach is unconventional and might seem exotic from the point of view of the conventional complex Hilbert space approach, however there are strong indications that it suits in many respects the effective description of entanglement. Here we shall restrict ourselves to the case of the pure state entanglement. We want to be selfcontained, therefore, we give necessary η-formalism illustrated by many examples and connections to the conventional description in Hilbert space. The factorization (separability) conditions and "measures" of entanglement for multiqubit systems are discussed for $n = 2, 3, 4$, in terms of the η-formalism.

2. Qubits and nilpotent commuting variables

As it is discussed in the Ref. 1, the nilpotent commuting variables are present in physical theories since late Eighties of the Twentieth Century. It seems that they were firstly used by Ziegler[2] in a model describing thermal fluctuations of the flux lattice to realize condition that flux lines cannot intersect or touch each other (so called hard-line interaction). Then this approach had been developed in series of works[3–5] among others to description of systems of hard core bosons[6] and applied for systems of macromolecules.[7] Slightly later, in 1993, such variables were independently introduced to quantum field theory by Palumbo[8] and then, based on them, a new approach of bosonization was successively developed (cf. Ref. 9–15 and recently 16,17).

In the approaches mentioned above: the first in statistical physics, the second in field theory, one is making use of the commuting nilpotent parameters and seem to be very promising. It turns out that the Lagrangian and Hamiltonian formulation of classical mechanics involving nilpotent commuting variables can be introduced.[23,26,27] One can consider classical limit for qubits, but at cost of introducing the mentioned above variables. Such a new approach is complementary to the conventional one, and gives natural setting for answering the entanglement questions. Some elements of such formalism were introduced in Ref. 18,28 (cf. Mandilara and Akulin contribution to this volume).

The one-particle space of the qubit is a two dimensional. The Fock space the tensor product of qubit states can be taken symmetric (cf. in this context Ref. 29). Qubit from this point of view exhibits mixture of the boson and fermion properties. To adequately describe qubits we have to play with the commutation or anticommutation relations on the one side, and symmetry properties of the tensor product on the other side. Bosons and fermions can be organized in unique graded structure where commutators, anticommutators, parity of elements of graded algebra, and symmetry of tensor product is consistent. They play distinguished role, because they describe fundamental particles. However, there are other useful objects related to parastatistics. Parafermions and parabosons were defined by tri-linear relations in Ref. 20 over fifty years ago and have their place in quantum field theory.[21]

Let us mention here the discussion of the definition of the qubit given by Wu and Lidar,[22] where they stress that qubits are neither bosons nor fermions, (in particular they consider realization of a qubit as a composite of fermions described by products of anticommuting variables). We refer to their definition of qubit as a parafermion and reformulate defining conditions using the commutator and nilpotency conditions. Such approach gives us natural realization of the relevant algebra, in analogy to the realization of the canonical anticommutation relations by means of the anticommuting variables and superdifferential operators.

The two dimensional state space of a single qubit system is naturally obtained from nilpotent (or more precisely, first order nilpotent i.e. with vanishing square) creation/anihilation operators. Taking commutators and symmetric tensor products for many-particle system we obtain parafermions, cf. Ref. 22. To find another description of qubits let us take the set of (anti)commutation relations for parafermions in the following form

$$[d_i, d_j^+]_- = \delta_{ij}(1 - 2N), \qquad [d_i, d_j]_- = 0, \qquad [d_i^+, d_j^+]_- = 0, \qquad (1)$$
$$d_i^2 = 0, \qquad (d_i^+)^2 = 0, \qquad (2)$$

where N can be seen as particle number operator. As it is known to obtain a classical limit for fermions, the introduction of the Grassmannian variables is necessary. Analogously for qubits, to get nontrivial classical limit we have to introduce nilpotent, but commuting variables.[1] Namely,

$$\eta\eta' = \eta'\eta, \qquad \eta^2 = \eta'^2 = 0 \tag{3}$$

The first order nilpotency encodes information that single qubit system is a two-level one, in the same time it is boson-like when considered in multi-qubit system.

The system of many qubits in the η-formalism has the same property as observed in[22] for parafermions, that it behaves for large n like boson, in paricular becames not nilpotent.

3. Canonical qubit relations

Understanding a qubit as a two-level quantum object exhibiting boson-like behavior in many particle system, and in the same time, being the subject of the Pauli exclusion principle. Such a hybrid object is not a fundamental particle cf. Ref. 8–10,12,18,22,23,26,27, like boson or fermion and inherently carries properties of a composed object, but now it is described without any explicit reference to the constituents and the way it is composed of.

To describe qubit as an object which is nilpotent but otherwise boson-like, let us observe that the commutator of qubit creation operator d^+ and qubit anihilation operator d if nontrivial, cannot have a value in the center of commutator algebra. This would give an contradiction with nilpotency. Making the following ansatz

$$[d, d^+]_- = 1 - 2N \tag{4}$$

and taking the compatibility condition with the nilpotency of d, d^+ one gets that

$$[d, d^+]_+ = 1 + 2Z, \tag{5}$$

where $Z = (d^+d - N)$ is an element from the center of the algebra. Moreover

$$[N, d]_+ = d \tag{6}$$

$$[N, d^+]_+ = d^+ \tag{7}$$

The canonical qubit relations (CQR) we shall consider consist of the following set of relations

$$[d, d^+]_- = 1 - 2N \tag{8}$$

$$[N, d]_- = d \tag{9}$$

$$[N, d^+]_- = -d^+ \tag{10}$$

$$d^2 = 0 = (d^+)^2 \tag{11}$$

Simplest realization we obtain for $N = d^+ d$. Moreover

$$(d + d^+)^2 = 1 + 2Z, \tag{12}$$

where for the $N = d^+ d$, $Z = 0$. As we shall describe it below, the CQR can be naturally realized within the η-function space by means of η-differential and multiplication operators.

Using for the CQR the argument of Wu and Lidar ([22]) one can see that the set of n qubits for the large n behaves like a boson. Namely

$$b = \frac{1}{\sqrt{n}} \sum_{i=1}^n d_i \tag{13}$$

$$b^+ = \frac{1}{\sqrt{n}} \sum_{i=1}^n d_i^+ \tag{14}$$

then

$$[b, b^+]_- = 1 - \frac{2}{n} \sum_{i=1}^n N_i. \tag{15}$$

Hence, when the number of particles n is much larger then the effective number of qubits we get $[b, b^+]_- \approx 1$.

4. Functions of nilpotent variables

For the description of multiqubit systems we shall consider set of functions of nilpotent commuting variables η. A number of variables is directly linked to the number of qubits we want to describe. We shall assume that variables η_i are first order nilpotent $(\eta^i)^2 = 0$, $\forall i$, and are algebraically independent i.e.

$$\eta^1 \cdot \eta^2 \cdot \ldots \eta^n \neq 0 \tag{16}$$

For further convenience we write $\vec{\eta} = (\eta^1, \eta^2, \ldots \eta^n)$ and the I_k denotes a strictly ordered multi-index $I_k = (i_1, i_2, i_3, \ldots, i_k)$. The function $f(\vec{\eta}) \in \mathcal{F}[\vec{\eta}]$

of n η-variables is defined by the following expansion

$$F(\vec{\eta}) = \sum_{k=0,I_k}^{n} F_{I_k} \eta^{I_k}, \qquad (17)$$

where $F_{I_k} \in \mathcal{N}$ are constant elements. The expansion (17) gives explicitly the dependence of a function F on the η-variables . The function F can also depend on the conventional real or complex variables $x \in \mathcal{K}^n$. In such case $F_{I_k}(x)$ encode this dependence. For example, for $n = 2$ we have

$$F(x,\vec{\eta}) = F(\eta_1, \eta_2) = F_0(x) + F_1(x)\eta_1 + F_2(x)\eta_2 + F_{12}(x)\eta_1\eta_2 \qquad (18)$$

It is instructive to consider the examples of elementary functions of η variables: power function, exponent, logarithm, trigonometric and hyperbolic functions. The explicit form of power function reveals interesting combinatorial structure of relevant terms exhibiting in their multiplicities. Summations are understood over strictly ordered configurations of indices.

n = 1:

$$F(\eta) = F_0 + F_1\eta \qquad (19)$$

$$F(\eta)^m = F_0^{m-1}(F_0 + F_1\eta), \qquad (20)$$

n = 2:

$$F(\eta^1, \eta^2) = F_0 + F_1\eta^1 + F_2\eta^2 + F_{12}\eta^1\eta^2 = F_0 + F_i\eta^i + F_{ij}\eta^i\eta^j \qquad (21)$$

$$F(\eta^1, \eta^2)^m = F_0^{m-1}(F_0 + mF^i\eta_i) + F_0^{m-2}(nF_0F_{12} \\ +m(m-1)F_1F_2)\eta_1\eta_2 \qquad (22)$$

n = 3:

$$\begin{aligned} F(\eta^1, \eta^2, \eta^3) &= F_0 + F_1\eta^1 + F_1\eta^2 + F_1\eta^3 + F_{12}\eta^1\eta^2 \\ &\quad + F_{13}\eta^1\eta^3 + F_{23}\eta^2\eta^3 + F_{123}\eta^1\eta^2\eta^3 \\ &= F_0 + F_i\eta^i + F_{ij}\eta^i\eta^j + F_{ijk}\eta^i\eta^j\eta^k \end{aligned} \qquad (23)$$

$$F(\eta^1, \eta^2, \eta^3)^m = F_0^{m-1}(F_0 + mF_i\eta^i) + F_0^{m-2}(mF_0F_{12}$$
$$+ m(m-1)F_1F_2)\eta_1\eta_2 + F_0^{m-2}(mF_0F_{13}$$
$$+ m(m-1)F_1F_3)\eta_1\eta_3 + F_0^{m-2}(mF_0F_{23} \quad (24)$$
$$+ m(m-1)F_2F_3)\eta_2\eta_3 + F_0^{m-3}(m(m-1)$$
$$F_0(F_1F_{23} + F_2F_{13} + F_3F_{12}) + m(m-1)$$
$$(m-2)F_1F_2F_3 + mF_0^2F_{123})\eta^1\eta^2\eta^3$$

What can be written in a compact form

$$F(\eta^1, \eta^2, \eta^3)^m = F_0^{m-1}(F_0 + mF_i\eta^i) + F_0^{m-2}(mF_0F_{ij}$$
$$+ m(m-1)F_iF_j)\eta_i\eta_j + F_0^{m-3}(m(m-1)$$
$$F_0F_iF_{jk} + m(m-1)(m-2)F_iF_jF_k$$
$$+ mF_0^2F_{ijk})\eta^i\eta^j\eta^k \quad (25)$$

$$\mathbf{n = 4:}$$

$$F(\eta^1, \eta^2, \eta^3, \eta^4) = F_0 + F_i\eta^i + F_{ij}\eta^i\eta^j + F_{ijk}\eta^i\eta^j\eta^k + F_{ijkl}\eta^i\eta^j\eta^k\eta^l \quad (26)$$

$$F(\eta^1, \eta^2, \eta^3, \eta^4)^m = F_0^{m-1}(F_0 + mF_i\eta^i) + F_0^{m-2}(mF_0F_{ij}$$
$$+ m(m-1)F_iF_j)\eta_i\eta_j + F_0^{m-3}(mF_0^2F_{ijk}$$
$$+ m(m-1)F_0F_iF_{jk} + m(m-1)(m-2)$$
$$F_iF_jF_k)\eta^i\eta^j\eta^k + F_0^{m-4}(nF_0^3F_{ijkl}$$
$$+ m(m-1)F_0F_iF_{jkl} + m(m-1)F_0F_{ij}F_{kl}$$
$$+ m(m-1)(m-2)(m-3)F_iF_jF_kF_l)$$
$$\eta^i\eta^j\eta^k\eta^l \quad (27)$$

To put the above formulas for powers of arbitrary $F(\vec{\eta})$ in a compact form we have introduced the following conventions: *(i)* when term gets negative power of $(F_0)^{m-k}$ - it vanishes; *(ii)* for the terms with the factor $(F_0)^0$ in front, we put $(F_0)^0 = 1$.

The η-exponent is defined using expansion analogous to the conventional one, what with the use of above power formulas gives

$$e^{F(\vec{\eta})} = \sum_{n=0}^{\infty} \frac{F(\vec{\eta})^n}{n!} = e^{F_0}e^{s(F(\vec{\eta}))} \quad (28)$$

Similarly, the definition of logarithm function is analogous to the conventional one. For η-functions with unit first term one can define

$$ln(1 + s(F(\vec{\eta})) = \sum_{k=1}^{\infty}(-1)^{k-1}\frac{s(F(\vec{\eta}))^k}{k}, \qquad (29)$$

where $s(F(\vec{\eta}))$ is the so called soul of a function $F(\vec{\eta})$ i.e. $F(\vec{\eta}) = 1 + s(F(\vec{\eta}))$. Using expansions with $F_0 = 1$ we obtain explicit expressions for $n = 1, 2, 3, 4$

n = 1:

$$ln(1 + s(F(\eta)) = F_1\eta \qquad (30)$$

n = 2:

$$ln(1 + s(F(\eta^1, \eta^2)) = F_i\eta^i + (F_{ij} - F_iF_j)\eta^i\eta^j \qquad (31)$$

n = 3:

$$ln(1 + s(F(\eta^1, \eta^2, \eta^3)) = F_i\eta^i + (F_{ij} - F_iF_j)\eta^i\eta^j + (F_{ijk} \qquad (32)$$
$$-F_iF_{jk} + 2F_iF_jF_k)\eta^i\eta^j\eta^k$$

n = 4:

$$ln(1 + s(F(\eta^1, \eta^2, \eta^3, \eta^4)) = F_i\eta^i + (F_{ij} - F_iF_j)\eta^i\eta^j + (F_{ijk} - F_iF_{jk}$$
$$+2F_iF_jF_k)\eta^i\eta^j\eta^k + (F_{ijkl} - F_{ij}F_{kl}$$
$$-F_iF_{jkl} + 2F_iF_jF_{kl} - 6F_iF_jF_kF_l)$$
$$\eta^i\eta^j\eta^k\eta^l \qquad (33)$$

From the point of view of the multiqubit entanglement a very interesting class of η-functions consists of the η-trigonometric functions. let $F = \sum F_{I_k}\eta^{I_k}$, $F_{I_k} \in \mathcal{R}$

$$cos(F(\vec{\eta})) = \sum_{k=0}(-1)^k\frac{F^{2k}}{(2k)!} \qquad (34)$$

$$sin(F(\vec{\eta})) = \sum_{k=0}(-1)^k\frac{F^{2k+1}}{(2k + 1)!} \qquad (35)$$

As in the conventional case there are valid various identities and reduction formulas e.g. $cos^2(F) + sin^2(F) = 1$. In particular

$$cos(F) = cos(F_0)cos(s(F)) - sin(F_0)sin(s(F)) \qquad (36)$$
$$sin(F) = sin(F_0)cos(s(F)) + cos(F_0)sin(s(F)). \qquad (37)$$

For further use let us write $sin(\sum \eta^i)$ and $cos(\sum \eta^i)$ in explicit form for $n=2,3,4$.

n = 2:

$$cos(\eta^1 + \eta^2) = 1 - \eta^1\eta^2 \tag{38}$$

$$sin(\eta^1 + \eta^2) = \eta^1 + \eta^2 \tag{39}$$

n = 3:

$$cos(\eta^1 + \eta^2 + \eta^3) = 1 - \eta^1\eta^2 - \eta^1\eta^3 - \eta^2\eta^3 \tag{40}$$

$$sin(\eta^1 + \eta^2 + \eta^3) = \eta^1 + \eta^2 + \eta^3 - \eta^1\eta^2\eta^3 \tag{41}$$

d = 4:

$$cos(\eta^1 + \eta^2 + \eta^3 + \eta^4) = 1 - \eta^1\eta^2 - \eta^1\eta^3 - \eta^1\eta^4 - \eta^2\eta^3 - \eta^2\eta^4$$
$$-\eta^3\eta^4 + \eta^1\eta^2\eta^3\eta^4 \tag{42}$$

$$sin(\eta^1 + \eta^2 + \eta^3 + \eta^4) = \eta^1 + \eta^2 + \eta^3 + \eta^4 - \eta^1\eta^2\eta^3 - \eta^1\eta^2\eta^4$$
$$-\eta^1\eta^3\eta^4 - \eta^2\eta^3\eta^4 \tag{43}$$

In the section 5 we shall bring explicit entangled states represented by η-trigonometric functions. To study a factorization problems it is convenient to have at hand the notion of derivative for functions of the η-variables. In the conventional case of functions of real or complex variables fundamental criteria of factorability make essential use of derivative of a function. The η-derivative is defined in the following way

$$\partial_i \eta^j = \delta_i^j, \quad \partial_i 1 = 0, \tag{44}$$

where

$$\partial_j = \frac{\partial}{\partial \eta^j} \tag{45}$$

then it is extended to the all functions by linearity i.e. $F[\vec{\eta}]$ i.e. $\partial_i(a\eta^k + b\eta^j) = a\partial_i\eta^k + b\partial_i\eta^j$. Obviously, $\partial_i\partial_j = \partial_j\partial_i$. Despite the simplicity of the above definition, the η-derivative is not a subject of the conventional Leibniz, instead for $F(\vec{\eta})$, $G(\vec{\eta})$ we have the following relation

$$\partial_i(F \cdot G) = \partial_i F \cdot G + F \cdot \partial_i G - 2\eta_i \partial_i F \partial_i G \tag{46}$$

This is an example of generalized Leibniz rule with the anomalous term. The following relations are direct consequence of the Eq. (46):

$$\partial_i(\eta_i F) = F - \eta_i \partial_i F \tag{47}$$

$$[\partial_i, \eta_i]_- = 1 - 2\eta_i \partial_i, \quad [\partial_i, \eta_i]_+ = 1 \tag{48}$$

For the functions of the nilpotent commuting variables one can also introduce the notion of the η-integral as follows

$$\int \eta_i d\eta_j = \delta_{ij}, \quad \int d\eta_i = 0 \qquad (49)$$

and by linearity extend it to all functions $F[\vec{\eta}]$. Such η-integral has the following properties:

$$\int \vec{\eta} d\vec{\eta} = 1, \quad \vec{\eta} = \eta_1 \eta_2 \ldots \eta_n, \quad d\vec{\eta} = d\eta_1 d\eta_2 \ldots d\eta_n \qquad (50)$$

$$\int \partial_i F(\vec{\eta}) d\eta_i = 0, \quad \text{and} \quad \int \partial_i F(\vec{\eta}) d\vec{\eta} = 0, \qquad (51)$$

where $F(\vec{\eta}) = F(\eta_1, \eta_2, \ldots, \eta_n)$ The integration by part formula for abouve integral, due to the modified Leibniz roule, has the following unconventional form

$$\left(\int F d\eta_i \right) \left(\int G \, d\eta_i \right) = \frac{1}{2} \left(\int (\partial_i F) \cdot G d\eta_i + \int F \cdot (\partial_i G) d\eta_i \right). \qquad (52)$$

Some changes of η-variables are allowed i.e. let matrix A represents permutation and scaling transformation, $\vec{\eta} = A\vec{\eta}'$ then

$$\int F(\vec{\eta}) d\vec{\eta} = (Per\, A)^{-1} \int F(A\vec{\eta}') d\vec{\eta}', \qquad (53)$$

where $Per\, A$ is the permanent of the matrix A. Its presence is characteristic for the η-formalism (as well as presence of the Hafnians).[a]

Finally, let us write a resolution of the η-generalization of the Dirac δ-function. Namely,

$$\int F(\vec{\eta}) \delta(\vec{\eta} - \vec{\rho}) d\vec{\eta} = F(\vec{\rho}) \qquad (54)$$

has the following resolution

$$\delta(\vec{\eta} - \vec{\rho}) = \prod_{i=1}^{n} (\eta_i + \rho_i), \qquad (55)$$

where ρ is nilpotent commuting variable, $\rho^2 = 0$. Proofs and further exposition of η-differential and integral calculus can be found in.[23]

It is natural to compare above objects to analogous ones known from the superanalysis. While the definition of the η-derivative and the η-integral

---

[a]Let us note that in the case of supermathematics natural link is towards Pfaffians and Determinants.

is the same as these introduced by Berezin, but its further properties depend on the multiplication in the algebra of functions what yields some differences. Fundamental one, is the modified Leibniz roule which contains modification providing consistency with the nilpotency of commuting variables. It also shows that the whole construction is nontrivial.

5. Generalized Hilbert space

To describe multiqubit systems within the η-formalism we shall need a generalization of the notion of Hilbert space. Here let us consider particular construction mimicking the Hilbert space of square integrable functions, such a generalized space of the η-functions we shall call the η-Hilbert space $\mathcal{H}$. In the space of such functions of n η-variables $F[\vec{\eta}_n]$, we introduce weakly non-degenerated scalar product defined by means of η-integral, in the following form

$$< F,\, G >_\mathcal{N} = \int F^*(\vec{\eta}) G(\vec{\eta}) e^{<\vec{\eta}^*,\vec{\eta}>} \, d\vec{\eta}^* \, d\vec{\eta}, = \int F^*(\vec{\eta}) G(\vec{\eta}) d\mu(\vec{\eta}^*,\, \vec{\eta}) \quad (56)$$

where

$$F^*(\vec{\eta}) = \sum_{k=0}^{n} \sum_{I_k} F_{I_k}^* \eta^{I_k\,*} \quad (57)$$

and $*$ denotes complex conjugation. For $F_0[\vec{\eta}_n]$ components F_{I_k} are complex number valued. The first order nilpotents η^{i*} are algebraically independent from η^i. In components we have

$$< F,\, G >_\mathcal{N} = \sum_{k=0}^{n} \sum_{I_k} F_{I_k}^* G_{I_k} \quad (58)$$

In particular, above η-scalar product has the following properties:

$$< \nu F, G > = < F, \nu^* G >, \quad \nu \in \mathcal{N} \quad (59)$$

$$< F, G > = 0, \quad \forall G \in \mathcal{H} \ \Rightarrow F = 0 \quad (60)$$

$$< F, G >^* = < G, F > \quad (61)$$

$$< F, F > \geq 0, \quad \forall F \in \mathcal{H} \quad (62)$$

We can represent in this formalism the 1-qubit algebra by taking $\mathcal{F}[\eta]$ (set of η-functions of one variable). Explicitly, the η-scalar product of $F(\eta)$ and $G(\eta)$ functions takes simple form

$$< F,\, G >_\mathcal{N} = F_0^* G_0 + F_1^* G_1. \quad (63)$$

Simple form of the realization of the qubit algebra is given in the following relations

$$d^+ = \eta \cdot , \qquad d = \partial/\partial\eta \tag{64}$$

We take the operators d^+ and d as conjugated with respect to the scalar product given by (58) moreover the operator

$$\sigma_3 = 1 - 2\eta\partial_\eta \tag{65}$$

is self-conjugated and $[d, d^+]_- = \sigma_3$. It is very interesting that in addition to the generalized scalar product there can be introduced in the n-qubit η-function space a weakly non-degenerate form with its symmetry properties depending on the parity of the qubit number n. Using the gradation mapping J we can consider the natural orthogonal projections on the even F_+ and odd F_- part of the function F, given in the following form

$$\pi_\pm F(\vec{\eta}) = \frac{1}{2}(F(\vec{\eta}) \pm J(F(\vec{\eta}))). \tag{66}$$

Then using the mapping J we can define the linear weakly non-degenerate form ω

$$\omega_n(F,\,G) = \int J(F(\vec{\eta})) \cdot G(\vec{\eta})d\eta_1 \ldots d\eta_n = \sum_{k,I_k}(-1)^k F_{I_k} G_{I_{n-k}}, \tag{67}$$

where $I_k \cup I_{n-k} = I_n$. The ω_n is symmetric or antisymmetric, depending on the parity of n

$$\omega_n(F,\,G) = (-1)^n\omega_n(G,\,F) \tag{68}$$

Such obtained form is the η-version of the form considered in the tensor product of $\mathbb{C}^2 \otimes \cdots \otimes \mathbb{C}^2$ from the antisymmetric form ϵ in the $\mathbb{C}^2$. For example, for $n = 1$ we have

$$\omega_1(F,\,G) = F_0 G_1 - F_1 G_0 \tag{69}$$

and for $n = 2$

$$\omega_2(F,\,G) = F_0 G_{12} + F_{12}G_0 - F_1 G_2 - F_2 G_1. \tag{70}$$

The last form written in particular basis $\{1, \eta_1,\, \eta_2,\, \eta_1\eta_2\}$ is given as follows[b]

$$\omega_2(1, \eta_1\eta_2) = 1, \quad \omega_2(\eta_1, \eta_2) = 1 \tag{71}$$

$$\omega_3(F,\,G) = F_0 G_{123} + F_{123}G_0 + F_{23}G_1 - F_1 G_{23}$$
$$+F_{13}G_2 - F_2 G_{13} + F_{12}G_3 - F_3 G_{12} \tag{72}$$

---

[b]It is equivalent to the form considered by Wallach.[32]

For further examples cf. Ref. 1.

There exists a natural family of projections related to the decomposition of the $F(\vec{\eta})$ into the part depending on the fixed η_k and independent of it. Namely,

$$F(\vec{\eta}) = F(\eta_1, \eta_2, \ldots, \hat{\eta}_k, \ldots, \eta_n) + \eta_k \tilde{F}(\eta_1, \eta_2, \ldots, \hat{\eta}_k, \ldots, \eta_n), \qquad (73)$$

where the hat indicates variable which is omitted. Natural realization of it is given by η-derivative operator, due to the identity

$$\partial_k \eta_k + \eta_k \partial_k = 1 \qquad (74)$$

we can introduce projectors

$$\pi_{k|0} = \partial_k \eta_k \cdot \qquad (75)$$

$$\pi_{k|1} = \eta_k \partial_k \cdot \qquad (76)$$

For a fixed k they are orthogonal and for different indices k they commute

$$\pi_{k|i} \pi_{k|j} = \delta_{ij} \pi_{k|j}, \quad \pi_{k|0} \oplus \pi_{k|1} = id \qquad (77)$$

$$\pi_{k|i} \pi_{l|j} = \pi_{l|j} \pi_{k|i}, \quad k \neq l \qquad (78)$$

For example the decomposition of the $F(\eta_1, \eta_2)$ has the following form

$$\begin{aligned} F(\eta_1, \eta_2) &= F_0 + F_2 \eta_2 + \eta_1(F_1 + F_{12}\eta_2) \equiv F(\eta_2) + \eta_1 \tilde{F}(\eta_2) \\ &= \pi_{1|0} F(\eta_1, \eta_2) + \pi_{1|1} F(\eta_1, \eta_2) \end{aligned} \qquad (79)$$

or

$$\begin{aligned} F(\eta_1, \eta_2) &= F_0 + F_1 \eta_1 + \eta_2(F_2 + F_{12}\eta_1) \equiv F(\eta_1) + \eta_2 \tilde{F}(\eta_1) \\ &= \pi_{2|0} F(\eta_1, \eta_2) + \pi_{2|1} F(\eta_1, \eta_2) \end{aligned} \qquad (80)$$

Moreover, we get the full decomposition of the $F(\eta_1, \eta_2)$ using composition of projectors

$$F_0 = \pi_{1|1} \pi_{2|0} F(\eta_1, \eta_2), \qquad (81)$$

$$F_1 \eta_1 = \pi_{1|1} \pi_{2|0} F(\eta_1, \eta_2), \qquad (82)$$

$$F_{12} \eta_1 \eta_2 = \pi_{1|1} \pi_{2|1} F(\eta_1, \eta_2) \qquad (83)$$

These formula generalize easily to the η-functions of n variables. Let us come back to the form ω. For $n = 1$ ω_1 is antisymmetric and can be realized in an alternative way by introducing the wedge product of η-functions as

$$F(\eta) \wedge G(\eta) \equiv (F_0 G_1 - F_1 G_0)\eta, \qquad (84)$$

Obtained in such a way an anti-symmetric form we shall denote by $\mathcal{D}_1$

$$\mathcal{D}_1(F, \ G) \equiv \int F(\eta) \wedge G(\eta) d\eta \tag{85}$$

Hence, for one η variable $\mathcal{D}_1(F, \ G) = \omega_1$. In the $d = 2$ case such generalization it is also possible in the following sense. The space of η-functions $F(\eta_1, \eta_2)$ is 2^2-dimensional, therefore it admits antisymmetric form. Because $d = 2$ η-function can be decomposed using one of the introduced above factorizations (79) or (80) therefore we can define $\mathcal{C}_2$ as

$$\mathcal{C}_2 \equiv \mathcal{D}_1(F(\eta_1), \tilde{F}(\eta_1)) = \mathcal{D}_1(F(\eta_2), \tilde{F}(\eta_2)) = F_0 F_{12} - F_1 F_2 \tag{86}$$

It can be recognized as a counterpart of the concurrence of the 2-qubit states. The anti-symmetric form $\mathcal{D}_2$ is defined in as follows

$$\begin{aligned}
\mathcal{D}_2 &\equiv \mathcal{D}_1(F(\eta_1), \ \tilde{G}(\eta_1)) - \mathcal{D}_1(G(\eta_1), \ \tilde{F}(\eta_1)) \\
&= \mathcal{D}_1(F(\eta_2), \ \tilde{G}(\eta_2)) - \mathcal{D}_1(G(\eta_2), \ \tilde{F}(\eta_2))
\end{aligned} \tag{87}$$

In the η-Hilbert spaces we can consider the simples possible basis composed of the monomials $\{\eta_{I_k}\}_{k=0}^n$. In such basis it is easy to set the correspondence to the so called computational basis in conventional Hilbert space approach widely used in the literature. The binary notation used there for describing the elements of the tensor product of $\mathbb{C}^2$ is simply related to our η-notation. Translation of the multi-index for system of n qubits with the "binary" entries 0, 1 to the multi-index I_k used in the η-function expansion is obtained by putting the ordinal numbers equal the position of 1's appearing in the binary multi-index e.g. $(0,0,0,0) \mapsto 0$, $(1,0,0,0) \mapsto 1$, $(0,1,0,0) \mapsto 2$, $\ldots$, $(0, 1, 0, 1) \mapsto (2,4)$, $\ldots$, $(1,1,1,1) \mapsto (1,2,3,4)$. In the case of two qubits we have that: $1 = |00>$, $\eta_1 = |10>$, $\eta_2 = |01>$ and $\eta_1\eta_2 = |11>$, and the trigonometric states get the following normalized form

$$\psi_{GHZ-} = \frac{1}{\sqrt{2}} cos(\eta_1 + \eta_2) = \frac{1}{\sqrt{2}}(|00> - |11>) \tag{88}$$

$$\psi_W = \frac{1}{\sqrt{2}} sin(\eta_1 + \eta_2) = \frac{1}{\sqrt{2}}(|01> + |10>) \tag{89}$$

Analogously for $n = 3$ qubit systems we have three η variables and trigonometrics functions give also the GHZ and W -states, and in addition emerges the cluster Wrener state. Namely,

$$cos(\eta^1 + \eta^2 + \eta^3) = 1 - \eta^1\eta^2 - \eta^1\eta^3 - \eta^2\eta^3 \tag{90}$$

$$sin(\eta^1 + \eta^2 + \eta^3) = \eta^1 + \eta^2 + \eta^3 - \eta^1\eta^2\eta^3 \tag{91}$$

what after normalization gives

$$\psi_{cos}^{(3)} = \frac{1}{2}cos(\eta^1 + \eta^2 + \eta^3) = \frac{1}{2}(1 - \eta^1\eta^2 - \eta^1\eta^3 - \eta^2\eta^3) \qquad (92)$$

$$= \frac{1}{2}(1 - \sqrt{3}\psi_{CW})$$

$$\psi_{sin}^{(3)} = \frac{1}{2}sin(\eta^1 + \eta^2 + \eta^3) = \frac{1}{2}(\eta^1 + \eta^2 + \eta^3 - \eta^1\eta^2\eta^3) \qquad (93)$$

$$= \frac{1}{2}(1 + \sqrt{3}\psi_W - \sqrt{2}\psi_{GHZ})$$

In the binary basis above states take the following form

$$\psi_{cos}^{(3)} = \frac{1}{2}(|000> - |110> - |101> - |011>) \qquad (94)$$

$$\psi_{sin}^{(3)} = \frac{1}{2}(|100> + |010> + |001> - |111>) \qquad (95)$$

and the cluster Werner state is defined as

$$\psi_{CW} = \frac{1}{\sqrt{3}}(\eta_1\eta_2 + \eta_1\eta_3 + \eta_2\eta_3) = \frac{1}{\sqrt{3}}(|110> + |101> + |011>). \qquad (96)$$

The bases composed of trigonometric η-functions are more interesting from the point of view of the entanglement. For the $n = 1$ such basis is identical with the monomial one $\{1, \eta\}$. For higher n we obtain interesting functions. For $n = 2$ we take trigonometric function (38) with arguments $\eta_1 \pm \eta_2$. These functions are orthogonal and we take them normalized with respect to our generalized scalar product

$$h_1 = \frac{1}{\sqrt{2}} cos(\eta^1 + \eta^2) = \frac{1}{\sqrt{2}}(1 - \eta^1\eta^2) = \psi_{GHZ-}, \qquad (97)$$

$$h_2 = \frac{1}{\sqrt{2}} cos(\eta^1 - \eta^2) = \frac{1}{\sqrt{2}}(1 + \eta^1\eta^2) = \psi_{GHZ+}, \qquad (98)$$

$$h_3 = \frac{1}{\sqrt{2}} sin(\eta^1 + \eta^2) = \frac{1}{\sqrt{2}}(\eta^1 + \eta^2) = \psi_{W+}, \qquad (99)$$

$$h_4 = \frac{1}{\sqrt{2}} sin(\eta^1 - \eta^2) = \frac{1}{\sqrt{2}}(\eta^1 - \eta^2) = \psi_{W-}. \qquad (100)$$

It is the η-realization of the "magic" basis for 2-qubit system. The cos-states are GHZ-type and sin-states are W-type (Bell states up to the particular phases). Above basis can be generalized to higher n, where it contains 2^n elements

$$\left\{ \frac{1}{\sqrt{2^{n-1}}} cos(\eta_1 \pm \eta_2 \cdots \pm \eta_n), \frac{1}{\sqrt{2^{n-1}}} sin(\eta_1 \pm \eta_2 \cdots \pm \eta_n) \right\} \qquad (101)$$

In our approach to description of qubits, as it was discussed in the Ref. 1, one introduce analog of the Schrödinger quantization procedure, which yields wave functions in the θ-variables, so called η-Schrödinger quantization, because we shall use the η-wavefunctions $\psi(x, \eta)$ and generalized Schrödinger equation, involving η-derivatives. For example, in the papers by Mandiliara at al.[18,28] was already used equation which can be named logarithmic η-Schrödinger equation. It was written for a so called tanglemeters, and used to address, via system control methods, some questions of the entanglement.

Let us recall, that there exists classical theory based on nilpotent commuting variables introduced in,[23] which provides configuration space description as well as the phase space description of nilpotent systems. It is called nilpotent classical mechanics. Moreover another essential aspect of the theory inviving nilpotent commuting variables i.e. path integral formalism, was discussed some time ago by Palumbo at al.[8–10] Let us note that in all this approaches, except the nilpotent classical mechanics there was neglected the fact, that the derivative with respect to the nilpotent commuting variables do not satisfy the Leibniz rule, what makes the whole construction nontrivial.

Therefore taking all above mentioned arguments into account, it is natural to consider the nilpotent quantum mechanics as formalism which is related by a "η-canonical quantization" to the classical nilpotent mechanics. Because known by now η-Poisson brackets do not satisfy the Jacobi identity, the term "η-canonical quantization" leaves some open questions, but the formalism of nilpotent quantum mechanics itself is consistent and very effective. We shall use here the restricted η-Schrödinger quantization in the following sense. To quantize classical nilpotent system, we take a classical observable in the normal ordered form i.e. momentum variables are to the right of the coordinate variables and realize position and momentum as operators

$$\eta_k \longrightarrow \hat{\eta}_k = \eta_k\cdot, \qquad p \longrightarrow \hat{p}_k = \frac{\partial}{\partial \eta_k} \tag{102}$$

in the $\mathcal{N}$-Hilbert space of η-functions depending on η_k, $k = 1, 2, \ldots, n$. Let $\psi \in \mathcal{F}[x, \vec{\eta}]$

$$i\hbar \frac{d}{dt} \psi(x, \vec{\eta}, t) = \hat{H} \psi(x, \vec{\eta}, t), \tag{103}$$

where $\hat{H}$ is quantized Hamiltonian $H(x, p_x, \eta, p_\eta, t)$ of the system. For the two level systems frequently one considers explicit time dependence of the

Hamiltonian. For example in the $n = 1$ case the Hamilton function is singular in the sense that its nilpotent part contains terms linear in p_η i.e. $H = \frac{1}{2m}p_x^2 + b(t)p_\eta + c(t)\eta p_\eta + V(x, \eta, t)$. After quantization we can write this Hamiltonian in the convenient form $\hat{H} = \frac{1}{2m}\hat{p}_x^2 + V(x) + \vec{B}(t)\cdot\vec{\sigma}$, where nilpotent part can be written as

$$\hat{H}_{nilp} = (B_x(t) + iB_y(t))\eta + (B_x(t) - iB_y(t))\frac{\partial}{\partial\eta} - 2B_z(t)\eta\frac{\partial}{\partial\eta} + B_z. \quad (104)$$

Considering only the nilpotent part (neglecting the simultaneous x coordinate dependence) one can assume the global factorization of time dependence of the η - wavefunction $\psi(\vec{\eta}, t)$ and study the stationary η-Schrödinger equation for nilpotent quantum system.

$$\hat{H}\psi(\vec{\eta}) = \lambda\psi(\vec{\eta}) \quad (105)$$

The structure of eigenstates for such multiqubit systems turns out to be nontrivial in view of the entanglement, cf. Ref. 1.

6. Factorization and entanglement chracterization in terms of η-functions determinants

The questions concerning existence of the entanglement can be answered using the criteria of factorization of η-functions using the η-differential calculus in some analogy to the factorization theory of functions of real or complex variable. As an introductory example let us consider the simple problem of linear independence of functions of one η-variable, which can be interpreted as special case of factorization

$$aF(\eta) + bG(\eta) = 0 \Leftrightarrow F(\eta) = -\frac{b}{a}G(\eta), \ a \neq 0.$$

As it is known[24] functions $F(\eta)$ and $G(\eta)$ are linearly dependent iff the Wronskian of the following matrix

$$W = \begin{pmatrix} F(\eta) & G(\eta) \\ \partial F(\eta) & \partial G(\eta) \end{pmatrix}, \quad (106)$$

$w = F_0 G_1 - F_1 G_0$ vanishes. In particular functions with nonvanishing first term in the expansion, i.e. $F_0 \neq 0$ can be written as $F(\eta) = F_0 e^{\frac{F_1}{F_0}\eta}$.

6.1. Factorization and entanglement measures of $F(\eta_1, \eta_2)$

To diagnose the possibility of factorization of the η-function of two variables one introduces[24] the determinant w_{12} of the following η-Wronski matrix

$$w_{12}(F(\eta_1, \eta_2)) = detW_{12} = \begin{vmatrix} F & \frac{\partial F}{\partial \eta_1} \\ \frac{\partial F}{\partial \eta_2} & \frac{\partial^2 F}{\partial \eta_1 \partial \eta_2} \end{vmatrix} = \begin{vmatrix} F & \partial_1 F \\ \partial_2 F & \partial_{12} F \end{vmatrix} = F_0 F_{12} - F_1 F_2 \tag{107}$$

For arbitrary function $F(\eta_1, \eta_2)$, $w_{12}(F) = 0$ if and only if $F(\eta_1, \eta_2) = G(\eta_1)\tilde{G}(\eta_2)$, for some G and $\tilde{G}$. Note that above Wronskian for the function of two η variables has numerical values (there is no explicit η dependence) and $w_{12} = \mathcal{H}$ (cf. Eq. 108). Using above formula let us introduce family of invariants $\mathcal{H}$ for various numbers of qubits $n = 2, 3, 4, \ldots$, defined as follows

$$\mathcal{H} = \sum_{k=0}^{[\frac{n}{2}]} \sum_{I_k} (-1)^k \left(\partial_{I_k} F(\vec{\eta}) \partial_{I_{n-k}} F(\vec{\eta}) \right) |_{\vec{\eta}=0} \tag{108}$$

where $\partial_{I_0} F(\vec{\eta}) = \partial_{\emptyset} F(\vec{\eta}) = F(\vec{\eta})$. In particular

$$n = 2:$$

$$\mathcal{H} = F_0 F_{12} - F_1 F_2 = w_{12}, \tag{109}$$

$$n = 3:$$

$$\mathcal{H} = F_0 F_{123} - F_1 F_{23} - F_2 F_{13} - F_3 F_{12}, \tag{110}$$

$$n = 4:$$

$$\mathcal{H} = F_0 F_{1234} - F_1 F_{234} - F_2 F_{134} - F_3 F_{124} - F_4 F_{123}$$
$$+ F_{12} F_{34} + F_{13} F_{24} + F_{14} F_{23} \tag{111}$$

In the above notation we do not distinguish $\mathcal{H}$ for variuos values of the n, but it will be clear from the context which one is under consideration. Examples:

(i) the Werner state represented by the function $\psi_W(\eta_1, \eta_2) = \frac{1}{\sqrt{2}}(\eta_1 + \eta_2)$; $w_{12}(\psi_W) = -\frac{1}{2}$ and indicates nonfactorability.

(ii) the GHZ state $\psi_{GHZ}(\eta_1, \eta_2) = \frac{1}{\sqrt{2}}(1 + \eta_1 \eta_2)$; $w_{12}(\psi_{GHZ}) = \frac{1}{2}$ and the η-function is nonfactorable,

(iii) the η-function $\psi = \frac{1}{2}(1 + \eta_1 + \eta_2 + \eta_1 \eta_2)$; $w_{12}(\psi) = 0$ and $\psi = \frac{1}{2}(1 + \eta_1)(1 + \eta_2)$.

The condition of vanishing Wronskian allows to distinguish several types of factorization: $e^{\eta_1}e^{\eta_2}$, $\eta_i e^{\eta_i}$, $\eta_i e^{\eta_j}$, where $i \neq j$; $i, j = 1, 2$. On the other hand the non-vanishing Wronskian gives two types of η-functions: GHZ-like ($e^{\pm \eta_1 \eta_2}$, $1 \pm \eta_i e^{\pm \eta_j}$, $e^{\pm \eta_1 \eta_2} \pm e^{\eta_i}$) and W-like ($1 \pm \eta_1 \pm \eta_2$, $\pm \eta_1 \pm \eta_2$). For further details see Ref. 1.

For two qubits the well known entanglement monotone is the concurrence. It can be expressed using the above defined Wronskian

$$\mathcal{C}(F(\eta_1, \eta_2)) = 2|w_{12}(F(\eta_1, \eta_2))|, \qquad < F, F >= 1, \qquad (112)$$

where the scalar product is defined in the generalized η-Hilbert space of η-functions $F(\eta_1, \eta_2)$ and in components it takes the form

$$< F, F >= |F_0|^2 + |F_0|^2 + |F_1|^2 + |F_2|^2 + |F_{12}|^2. \qquad (113)$$

Using the notion of the comb[48] and antilinear mapping $F \mapsto F^c$, where $F^c = (\sigma^y \otimes \sigma^y)\bar{F}$ one can express concurrence as

$$\mathcal{C}(F) = | < F^c, F > | = 2|F_0 F_{12} - F_1 F_2|. \qquad (114)$$

The operator $\sigma^y \otimes \sigma^y$ is realized in the following form

$$\sigma^y \otimes \sigma^y = -(\partial_1 \partial_2 + \eta_1 \eta_2 - \eta_2 \partial_1 - \eta_1 \partial_2) \qquad (115)$$

As it is shown,[24] by taking modulus of the scalar product of the entries of the Wronski matrix we can write the so called visibility $\mathcal{V}_i$ of ith qubit in terms of an η-function

$$\mathcal{V}_i = 2| < \partial_i F, F > | \qquad (116)$$

and, on the other hand, the predictability $\mathcal{P}_i$ for each qubit can also be expressed in a natural way using a reflection of relevant η-variable i.e.

$$\mathcal{P}_i = | < F, J_i(F) > |, \qquad (117)$$

where $J_1(F(\eta_1, \eta_2)) = F(-\eta_1, \eta_2)$ and $J_2(F(\eta_1, \eta_2)) = F(\eta_1, -\eta_2)$. Finally we can formulate for η-functions the so called complementarity relations for system of two qubits

$$\mathcal{C}^2(F) + \mathcal{V}_i^2(F) + \mathcal{P}_i^2(F) =< F, F >^2= 1, \quad i = 1, 2 \qquad (118)$$

6.2. Factorization and entanglement measures of $F(\eta_1, \eta_2, \eta_3)$

In the case of three variables we have to consider a set of the Wronski matrices, for all distinct pairs of variables. From such a bipartite information one can determine the level of non/factorability of $F(\eta_1, \eta_2, \eta_3)$. This time the Wronskians depend explicitly on η-variables

$$
\begin{aligned}
w_{12}(F)(\eta_3) &= w_{12}(F|_{\eta_3=0}) + (\mathcal{H} + 2F_3F_{12})\eta_3 \\
&\equiv w_{12}(F|_{\eta_3=0}) + \tilde{\mathcal{H}}_3\eta_3
\end{aligned} \tag{119}
$$

$$
\begin{aligned}
w_{13}(F)(\eta_2) &= w_{13}(F|_{\eta_2=0}) + (\mathcal{H} + 2F_2F_{13})\eta_2 \\
&\equiv w_{13}(F|_{\eta_2=0}) + \tilde{\mathcal{H}}_2\eta_2
\end{aligned} \tag{120}
$$

$$
\begin{aligned}
w_{23}(F)(\eta_1) &= w_{23}(F|_{\eta_1=0}) + (\mathcal{H} + 2F_1F_{23})\eta_1 \\
&\equiv w_{23}(F|_{\eta_1=0}) + \tilde{\mathcal{H}}_1\eta_1,
\end{aligned} \tag{121}
$$

where $\mathcal{H}$ is given by (110) according to Eq.(108). What is interesting, one can express all terms in the above expansions of η-function determinants in terms of the invariants of Wronski matrices. Namely, we find that the $\tilde{\mathcal{H}}_k(F)$ can be written in terms of traces

$$
\tilde{\mathcal{H}}_k(F) = trW_{ij}(F|_{\eta_k=0}) \cdot trW_{ij}(\partial_k F) - tr\left(W_{ij}(F|_{\eta_k=0})W_{ij}(\partial_k F)\right) \tag{122}
$$

For the η-functions of three variables one can introduce effective criterion indicating factorization of two subsystems $(i-k)(j)$, with true separation of dependence on one variable (cf. Ref. 24). Let $F = F(\eta_1, \eta_2, \eta_3)$, there exist functions $G(\eta_i, \eta_k)$ and $\tilde{G}(\eta_j)$ such that $F(\eta_1, \eta_2, \eta_3) = G(\eta_i, \eta_k)\tilde{G}(\eta_j)$ iff the following conditions are satisfied: $w_{ij}(F)(\eta_k) = 0$, $w_{kj}(F)(\eta_i) = 0$, $w_{ij}(\partial_k F) = 0$, and $w_{kj}(\partial_i F) = 0$ $(i, j, k = 1, 2, 3,$ are all different and fixed). Let us note that when there exists the logarithm of the η-function then, condition for separability of variables takes very simple form.[18] For example the dependence on η_j variable is separable if and only if

$$
\partial_i\partial_j ln(F) = 0 = \partial_k\partial_j ln(F), \qquad i, j, k = 1, 2, 3. \tag{123}
$$

When above derivatives of the lnF vanishes, it equivalently means that: $w_{ik}(F) = 0 = w_{jk}(F)$ e.g. $\partial_1\partial_3 lnF(\eta_1, \eta_2, \eta_3) = 0 \Leftrightarrow (w_{13}(F|_{\eta_2}) = 0$ and $\tilde{\mathcal{H}}_2 = 0)$. In such case, the rest of the conditions present in the factorization criterion is fulfilled automatically. Such simple characterization of separability in terms of logarithm generalizes to higher n,[18] but one has to remember, that the set of functions with vanishing body $b(F) = 0$, and hence not having logarithm, is large and grows with the value of n. All Werner-like states fall into this set. The decomposition of $F(\vec{\eta})$ into factors is not unique in general. Let us note that factorization properties of the η-function $F(\eta_1, \eta_2, \eta_3)$ and its dual $\star F(\eta_1, \eta_2, \eta_3)$ are closely related, because

$$
w_{ij}(F|_{\eta_k=0}) = w_{ij}(\partial_k(\star F)), \qquad \tilde{\mathcal{H}}_i(F) = \tilde{\mathcal{H}}_i(\star F) \tag{124}
$$

To see how non/factorization properties of the $F(\eta_1, \eta_2, \eta_3)$ are encoded in values of relevant Wronskians let us consider some examples. For separability the normalization of the function is unimportant, but in view of application to entanglement monotones we shall consider normalized η-functions.

Examples:

(1) The Werner state is represented by the η-function $\psi_W = \frac{1}{\sqrt{3}}(\eta_1 + \eta_2 + \eta_3)$, it is non-factorable. We have

$$w_{ij}(\psi_W|_{\eta_k=0}) = -\frac{1}{3}, \qquad \tilde{\mathcal{H}}_i = 0, \tag{125}$$

$$w_{ij}(\partial_k \psi_W) = 0 \tag{126}$$

The cluster Werner state is represented by the η-function dual to ψ_W i.e. $\star\psi_W = \frac{1}{\sqrt{3}}(\eta_1\eta_2 + \eta_1\eta_3 + \eta_2\eta_3)$, as before it is non-factorable. Namely, according to observation concerning dual functions[24]

$$w_{ij}(\star\psi_W|_{\eta_k=0}) = 0, \qquad \tilde{\mathcal{H}}_i = 0, \tag{127}$$

$$w_{ij}(\partial_k \psi_W) = -\frac{1}{3} \tag{128}$$

Obviously get that $\star\psi_W = \frac{\sqrt{3}}{2}(\psi_W)^2$.

(2) The GHZ state: $\psi_{GHZ} = \frac{1}{\sqrt{2}}(1 + \eta_1\eta_2\eta_3)$. This function also is not factorable, because of relations

$$w_{ij}(\psi_{GHZ}|_{\eta_k=0}) = 0, \qquad \tilde{\mathcal{H}}_k = \frac{1}{2} \tag{129}$$

$$w_{ij}(\partial_k \psi_{GHZ}) = 0, \tag{130}$$

We have that $\psi_{GHZ} = \star\psi_{GHZ}$. The nontrivial contribution to the Wronskian η-function comes here only from the $\tilde{H}$ i.e. $w_{ij}(\psi_{GHZ}) = \eta_k$.

(3) Let ψ be a factorable state of the form $\psi = \frac{1}{2}(1 + \eta_3 + \eta_1\eta_2 + \eta_1\eta_2\eta_3)$, then

$$w_{12}(\psi|_{\eta_3=0}) = \frac{1}{4}, \qquad w_{12}(\partial_3\psi) = \frac{1}{4}, \qquad \tilde{\mathcal{H}}_3 = \frac{1}{2} \tag{131}$$

$$w_{13}(\psi|_{\eta_2=0}) = 0, \qquad w_{13}(\partial_2\psi) = 0, \qquad \tilde{\mathcal{H}}_2 = 0 \tag{132}$$

$$w_{23}(\psi|_{\eta_1=0}) = 0, \qquad w_{23}(\partial_1\psi) = 0, \qquad \tilde{\mathcal{H}}_1 = 0 \tag{133}$$

In this case $w_{i3}(\psi)$ and $w_{i3}(\partial_j\psi)$ vanish, so dependence on η_3 can be factorized and indeed $\psi \sim (1 + \eta_1\eta_2)(1 + \eta_3)$ is a product of the GHZ-function and e^{η_3}.

Presented classification of factorization of the functions for the $n = 3$ reveals[24] the already known onion structure in the space of states discussed in Ref. 44,45. As it is known, one can distinguish three sets B_i of mutually bipartite separable functions with nonempty common part of totaly separable functions. Descriptively, the B_i form a rosette which is surrounded by the set B_W of nonseparable functions of W-type nonseparability (as illustrated in above examples) and of the set B_{GHZ} of nonseparable functions of GHZ-type. Counting dimensions of our η-function spaces, we take normalized functions, moreover, when function is separable we normalize each factor independently. Resulting dimensions of above sets are the following

$$\dim(B_i) = 3 + 1 = 4 \tag{134}$$

$$\dim(\cap_i B_i) = 1 + 1 + 1 = 3 \tag{135}$$

$$\dim(B_W) = 2^3 - 1 - 1 = 6 \tag{136}$$

$$\dim(B_{GHZ}) = 2^3 - 1 = 7 \tag{137}$$

where for the B_W we have taken into account that there vanishes component F_0 of the W-type function. These dimensions agree with the results obtained from the invariants theory.[45]

The principal entanglement monotone used to detect and measure the "degree" of non-separability of 3-qubit systems is the 3-tangle.[50,51] Mathematically it is based on the hyperdeterminant known in invariants theory for a long time

$$\tau_{123} = 4|Det(F)| \tag{138}$$

Where hyperdeterminant realized in terms of components of the η-function $F(\eta_1, \eta_2, \eta_3)$ has the following form

$$\begin{aligned}
Det(F) = {} & (F_0^2 F_{123}^2 + F_3^2 F_{12}^2 + F_2^2 F_{13}^2 + F_1^2 F_{23}^2) + 4(F_0 F_{23} F_{13} F_{12} \\
& + F_1 F_2 F_3 F_{123}) - 2(F_0 F_3 F_{12} F_{123} + F_0 F_2 F_{13} F_{123} \\
& + F_0 F_1 F_{23} F_{123} + F_2 F_3 F_{13} F_{12} + F_1 F_3 F_{23} F_{12} \\
& + F_1 F_2 F_{23} F_{13})
\end{aligned} \tag{139}$$

For the GHZ state it gives maximal value, but it do not show the entangled character of the Werner state. As Coffman Kundu and Wooters have shown, the 3-tange τ_{123} and mutual concurrences of bipartite systems of three qubits (we number qubits instead labeling them by the letter) are subject of the following relation

$$C_{1(23)}^2 = C_{12}^2 + C_{13}^2 + \tau_{123} \tag{140}$$

Using this equation, one can define symmetric entanglement monotone symmetric, averaged over all configurations of qubits. Namely,

$$Q(F) = \frac{1}{3}(\mathcal{C}^2_{1(23)} + \mathcal{C}^2_{2(13)} + \mathcal{C}^2_{3(12)}) = \frac{2}{3}(\mathcal{C}^2_{12} + \mathcal{C}^2_{13} + \mathcal{C}^2_{23}) + \tau_{123} \quad (141)$$

Such Q is the $n = 3$ realization of global entanglement measure introduced by Meyer and Wallach[36] for arbitrary n and also considered by Brennen[52] with slightly different form of this function. As it was shown in the Ref. 24 hyperdeterminant for the η-function can be expressed as

$$Det(F) = \frac{1}{3}\sum_k (\tilde{\mathcal{H}}^2_k - 4w_{ij}(F|_{\eta_k=0})w_{ij}(\partial_k F), \quad i \neq j \neq k \quad (142)$$

With the use of the η-realization of σ matrices and the η-scalar product, above relation takes the following form

$$| < \sigma^{ij}_2 \bar{F}, F > | = 2|w_{ij}(F|_{\eta_k=0}) + w_{ij}(\partial_k F)|, \quad (143)$$

where bar denotes complex conjugation and σ^{ij}_2 is a tensor product of $\mathbb{I}$ and σ_2 matrices on i^{th} and j^{th} positions e.g. $\sigma^{23}_2 = \mathbb{I} \otimes \sigma_2 \otimes \sigma_2$

$$\mathcal{C}_{ij} = 2|w_{ij}(F|_{\eta_k=0}) + w_{ij}(\partial_k F)| \quad (144)$$

$$Q(F) = \frac{2}{3}\sum_{i<j} | < \sigma^{ij}_2 \bar{F}, F > |^2 + \tau_{123} \quad (145)$$

$$Q(F) = \frac{4}{3}\left(2\sum_k |w_{ij}(F|_{\eta_k=0}) + w_{ij}(\partial_k F)|^2 \right.$$
$$\left. + |\sum_k(\tilde{\mathcal{H}}^2_k - 4w_{ij}(F|_{\eta_k=0})w_{ij}(\partial_k F)|)\right) \quad (146)$$

With the use of relation (122) the $Q(F)$ can be expressed solely in terms of determinants and traces of the Wronski matrices $W_{ij}(F)$, $W_{ij}(\partial_k F)$ namely,

$$Q(F) = \frac{4}{3}(\sum_k |w_{ij}(F|_{\eta_k=0}) + w_{ij}(\partial_k F)|^2 + |\sum_k((trW_{ij}(F|_{\eta_k=0})$$
$$trW_{ij}(\partial_k F) - tr(W_{ij}(F|_{\eta_k=0})W_{ij}(\partial_k F)))^2$$
$$-4w_{ij}(F|_{\eta_k=0})w_{ij}(\partial_k F)|). \quad (147)$$

Above representation shows explicitly how entanglement and factorability properties are intertwined. Let us test behavior of of above function on some states, where the normalization of η-function is important, to have fixed scale of values for Q.

Examples:

(i) $Q(\frac{1}{\sqrt{3}}(\eta_1 + \eta_2 + \eta_3)) = \frac{8}{9}$ The ψ_W state contributes to the value of Q only by the relative 2-qubit entanglement, the 3-tangle τ_{123} for this state vanishes. Analogously we get the same value for dual to ψ_W η-function $\star\psi_W$: $Q(\frac{1}{\sqrt{3}}(\eta_1\eta_2 + \eta_1\eta_3 + \eta_2\eta_3)) = \frac{8}{9}$

(ii) The GHZ state cotributes to the value of Q only through the 3-tangle $Q(\psi_{GHZ} = \frac{1}{\sqrt{2}}(1 + \eta_1\eta_2\eta_3)) = 1$

(iii) It is interesting to compare behavior of Q on states of the form $\sim 1 + \star\psi_W$ and $\sim 1 + \psi_W$. For the first one we get $Q(\frac{1}{2}(1 + \eta_1\eta_2 + \eta_1\eta_3 + \eta_2\eta_3)) = 1$ with contribution solely from 3-tangle, but for the second state we get only contribution from the relative 2-qubit entanglement and $Q(\frac{1}{2}(1 + \eta_1 + \eta_2 + \eta_3)) = \frac{3}{4}$.

6.3. *Factorization and entanglement measures of* $F(\eta_1, \eta_2, \eta_3, \eta_4)$

To detect the possibility factorization one can consider[24] family of Wronski 2×2 matrices. For functions of four η-variables the determinants $w_{ij}(F(\eta_1, \eta_2, \eta_3, \eta_4))$ are functions of still two η-variables. We have

$$w_{ij}(F) = w_{ij}(F|_{\eta_k=\eta_l=0}) + \tilde{\mathcal{H}}_k(F|_{\eta_l=0})\eta_k + \tilde{\mathcal{H}}_l(F|_{\eta_k=0})\eta_l + \tilde{\mathcal{H}}_{kl}(F)\eta_k\eta_l, \tag{148}$$

where e.g. $\tilde{\mathcal{H}}_{34} = \mathcal{H} + 2F_3F_{124} + 2F_4F_{123} - 2F_{13}F_{24} - 2F_{14}F_{23}$ and $\mathcal{H}$. It is the Cayley determinant given by Eq. (111). As before we use notation $\pi_i(F) = F|_{\eta_i}$.

The separability condition in the case of $n = 4$ can be formulated for two types of factorization:[24] (i-j)(k-l) and (i-j-k)(l)):

(i-j)(k-l): a function $F(\eta_1, \eta_2, \eta_3, \eta_4)$ is factorable into the product of functions depending on separated pairs of variables $F = G(\eta_{i_1}, \eta_{i_2})\tilde{G}(\eta_{j_1}, \eta_{j_2})$, $I = \{i_1, i_2\}$ and $J = \{j_1, j_2\}$ $I \cup J = \{1, 2, 3, 4\}$, where $i_1 < i_2$, $j_1 < j_2$ then,

$$w_{i_k j_l}(F) = 0, \tag{149}$$

$$w_{i_k j_l}(\partial_{i_{k'}} F) = 0, \tag{150}$$

$$w_{i_k j_l}(\partial_{j_{l'}} F) = 0, \tag{151}$$

$$w_{i_k j_l}(\partial_{i_{k'}}\partial_{j_{l'}} F) = 0, \tag{152}$$

where $i_k \neq i_{k'} \in I$ and $j_l \neq j_{l'} \in J$.

(i-j-k)(l): a function $F(\eta_1, \eta_2, \eta_3, \eta_4)$ has decomposition into the product $F = G(\eta_{i_1}, \eta_{i_2}, \eta_{i_3})\tilde{G}(\eta_j)$ then,

$$w_{i_k j}(F) = 0, \tag{153}$$

$$w_{i_k j}(\partial_{i_{k'}} F) = 0, \tag{154}$$

$$w_{i_k j}(\partial_{i_{k''}} F) = 0, \tag{155}$$

$$w_{i_k j}(\partial_{i_{k'} i_{k''}} F) = 0, \tag{156}$$

where $J = \{i_1, i_2, i_3\}$, $J = \{j\}$ and as before $I \cup J = \{1, 2, 3, 4\}$; $i_k \neq i'_k \neq i_{k''}$. For functions with nonvanishing first component F_0 there exists logarithm $ln(1 + s(F))$ and in such a case factorization conditions are:

$$\partial_{i_k} \partial_{j_l} ln(1 + s(F)) = 0, \quad k, l = 1, 2 \tag{157}$$

for factorizations of type $(i - j)(k - l)$ or

$$\partial_{i_k} \partial_j ln(1 + s(F)) = 0, \quad k, = 1, 2, 3 \tag{158}$$

for factorizations of type $(i - j - k)(l)$. We just get special cases of the analog of the d'Alembert criterion considered in Ref. 18. Let us stress that criterions written directly in terms of Wronskians of η-function F are more general and apply to all η-functions.

For four qubits we can introduce nontrivial 4×4 Wronski matrices of the following form

$$\mathcal{L}_{ij} = \begin{pmatrix} W_{ij}(F) & W_{ij}(\partial_k F) \\ W_{ij}(\partial_l F) & W_{ij}(\partial_l \partial_k F) \end{pmatrix}, \tag{159}$$

where (ij) and (lk) are ordered indices, hence $i < j$, $l < k$. The numerical part of these matrices we shall call the body. Such body can be written as conventional matrix with complex number's entries, in the following form

$$L_{ij} = \begin{pmatrix} W_{ij}(\pi_l \pi_k F) & W_{ij}(\pi_l \partial_k F) \\ W_{ij}(\pi_k \partial_l F) & W_{ij}(\partial_l \partial_k F) \end{pmatrix} \tag{160}$$

It turns out that above (L_{ij}) are related to the known in the literature matrices obtained from the classical invariants theory, namely using results of Ref. 46 we get

$$det\, L_{12} = N, \quad det\, L_{14} = M, \quad det\, L_{13}^{PT} = L, \tag{161}$$

where L_{13}^{PT} means partial transposition of the matrix L_{13}

$$L_{13}^{PT} = \begin{pmatrix} W_{13}(\pi_2 \pi_4 F) & W_{13}(\pi_4 \partial_2 F) \\ W_{13}(\pi_2 \partial_4 F) & W_{13}(\partial_2 \partial_4 F) \end{pmatrix}. \tag{162}$$

The presence of partial transposition is important here, because invariants L, M, N satisfy the relation[46]

$$L + N + M = 0 \qquad (163)$$

Such constraint for Wronski matrices (159) would be problematic. The η-function valued determinants of $\mathcal{L}_{ij}$ give information about possibility of factorization and we are interested in all matrices $\mathcal{L}_{ij}$, however for the complementary pairs of indices we get the same values of the determinant, so is enough to consider e.g.: L_{12}, L_{13}, L_{14}. While above conditions of separability for $n = 4$ are invertible, we can get much weaker ones using determinants of 4×4 matrices: $\mathcal{L}_{ij}$. It turns out that, a necessary condition for $F(\eta_1, \eta_2, \eta_3, \eta_4)$ to be factorable into the product $G(\eta_{i_1}, \eta_{i_2}, \eta_{i_3})\tilde{G}(\eta_j)$ is the following

$$det\,\mathcal{L}_{i_k j} = 0, \qquad k = 1, 2, 3 \qquad (164)$$

When F is factorable into a product $F = G(\eta_{i_1}, \eta_{i_2})\tilde{G}(\eta_{j_1}, \eta_{j_2})$ then

$$det\,\mathcal{L}^{PT}_{i_k j_l} = 0, \qquad k, l = 1, 2. \qquad (165)$$

Let us note that for this type of separation of variables we take partial transposition of $\mathcal{L}_{ij}$. As an illustration let us see the how is detected that W-state and GHZ-state are not factorable.

(i) For the $n = 4$ GHZ-state $\psi_{GHZ} = \frac{1}{\sqrt{2}}(1 + \eta_1\eta_2\eta_3\eta_4)$ we have

$$w_{ij}(\psi_{GHZ}) = \frac{1}{2}\eta_l\eta_k, \qquad (166)$$

$$w_{ij}(\partial_l\psi_{GHZ}) = 0 \qquad (167)$$

$$w_{ij}(\partial_k\partial_l\psi_{GHZ}) = 0 \qquad (168)$$

We see that here the body (numerical part) of w_{ij} vanishes i.e. $b(w_{ij}(\psi_{GHZ})) = 0$, and $\mathcal{H}(\psi_{GHZ}) = \frac{1}{2}$, but $det\,\mathcal{L}_{ij} = 0$ and also $M = N = L = 0$. The Wronskians w_{ij} detect the nonfactorability, and what is important, not only the body of the w_{ij} is important.

(ii) For the $n = 4$ W-state we have $\frac{1}{2}(\eta_1 + \eta_2 + \eta_3 + \eta_4)$. Here $\mathcal{H}(\psi_W) = 0$, and

$$w_{ij}(\psi_W) = -\frac{1}{4} \qquad (169)$$

$$w_{ij}(\partial_l\psi_W) = 0 \qquad (170)$$

$$w_{ij}(\partial_k\partial_l\psi_W) = 0. \qquad (171)$$

For the Werner type state the nonzero term of the w_{ij} is contained in the body of this Wronskian. As before: $det \, \mathcal{L}_{ij} = 0$ and $M = N = L = 0$.

Let us recall here that for $n = 4$ we have reach set of SLOCC inequivalent entangled states. Namely, there are nine equivallence classes. Classification was firstly obtained by Verstraete et al. in.[54] One introduces the normal forms to which a given state can be uniquely transformed. Such normal form can depend on some invariants of SLOCC transformations and in such case gives family of orbits parametrized by these invariants. There are also single orbits, related to the parameter independent normal forms. For example the family of orbits $|\Psi_1\rangle$ originally denoted in the Ref. 54 as G_{abcd} is parametrized by four invariants. Using the η-function approach it is easy to realize that the G_{abcd} can be written in the form[24]

$$G_{abcd} = \Psi_1 = \frac{a}{2}\psi_{GHZ+}^{(12)}\psi_{GHZ+}^{(34)} + \frac{d}{2}\psi_{GHZ-}^{(12)}\psi_{GHZ-}^{(34)}$$
$$+ \frac{b}{2}\psi_{W+}^{(12)}\psi_{W+}^{(34)} + \frac{c}{2}\psi_{W-}^{(12)}\psi_{W-}^{(34)} \tag{172}$$

In the following we shall use slightly modified normal forms, proposed by Chterental and Doković.[53] We present below the literal translation to the η-function form of representatives given by Chterental and Doković.[53] For convenience of the reader let us firstly recall the states in the form given in Ref. 53. Namely, in the binary basis they are given as follows

$$|\Psi_1\rangle = \frac{a+d}{2}(|0000\rangle + |1111\rangle) + \frac{a-d}{2}(|0011\rangle + |1100\rangle)$$
$$+ \frac{b+c}{2}(|0101\rangle + |1010\rangle) + \frac{b-c}{2}(|0110\rangle + |1001\rangle) \tag{173}$$

$$|\Psi_2\rangle = \frac{a+c-i}{2}(|0000\rangle + |1111\rangle) + \frac{a-c+i}{2}(|0011\rangle + |1100\rangle)$$
$$+ \frac{b+c+i}{2}(|0101\rangle + |1010\rangle) + \frac{b-c-i}{2}(|0110\rangle + |1001\rangle)$$
$$+ \frac{i}{2}(|0001\rangle + |0111\rangle + |1000\rangle + |1110\rangle$$
$$- |0010\rangle - |0100\rangle - |1011\rangle - |1101\rangle) \tag{174}$$

$$|\Psi_3\rangle = \frac{a}{2}(|0000\rangle + |1111\rangle + |0011\rangle + 1100\rangle) + \frac{b+1}{2}(|0101\rangle + |1010\rangle)$$
$$+ \frac{b-1}{2}(|0110\rangle + |1001\rangle) + \frac{1}{2}(|1101\rangle + |0010\rangle - |0001\rangle - |1110\rangle) \tag{175}$$

$$|\Psi_4\rangle = \frac{a+b}{2}(|0000\rangle + |1111\rangle) + b(|0101\rangle + |1010\rangle) + i(|1001\rangle - |0110\rangle)$$

$$+ \frac{a-b}{2}(|0011\rangle + |1100\rangle) + \frac{1}{2}(|0010\rangle + |0100\rangle + |1011\rangle + |1101\rangle$$

$$- |0001\rangle - |0111\rangle - |1000\rangle - |1110\rangle) \tag{176}$$

$$|\Psi_5\rangle = a(|0000\rangle + |0101\rangle + |1010\rangle + |1111\rangle) - 2i(|0100\rangle - |1001\rangle - |1110\rangle) \tag{177}$$

$$|\Psi_6\rangle = \frac{a+i}{2}(|0000\rangle + |1111\rangle + |0011\rangle + 1100\rangle) + \frac{a-i+1}{2}(|0101\rangle$$

$$+ |1010\rangle) + \frac{a-i-1}{2}(|0110\rangle + |1001\rangle) + \frac{i+1}{2}(|1101\rangle + |0010\rangle)$$

$$+ \frac{i-1}{2}(|0001\rangle + |1110\rangle) - \frac{i}{2}(|0100\rangle + |0111\rangle + |1000\rangle + |1011\rangle) \tag{178}$$

$$|\Psi_7\rangle = (|0101\rangle - |0110\rangle + |1100\rangle + |1111\rangle) + (i+1)(|1001\rangle + |1010\rangle)$$

$$- i(|0100\rangle + |0111\rangle + |1101\rangle - |1110\rangle) \tag{179}$$

$$|\Psi_8\rangle = \frac{i+1}{2}(|0000\rangle + |1111\rangle - |0010\rangle - |1101\rangle)$$

$$+ \frac{i-1}{2}(|0001\rangle + |1110\rangle - |0011\rangle - 1100\rangle)$$

$$+ \frac{1}{2}(|0100\rangle + |1001\rangle + |1010\rangle + |0111\rangle)$$

$$+ \frac{1-2i}{2}(|1000\rangle + |0101\rangle + |0110\rangle + |1011\rangle) \tag{180}$$

$$|\Psi_9\rangle = \frac{1}{2}(|0\rangle + |1\rangle) \otimes (|000\rangle + |011\rangle + |100\rangle + |111\rangle$$

$$+ i(|001\rangle + |010\rangle - |101\rangle - |110\rangle)) \tag{181}$$

It is instructive to see their literal ranslation into the η-function form

$$\Psi_1 = \frac{a+d}{2}e^{\vec{\eta}} + \frac{a-d}{2}(\eta_1\eta_2 + \eta_3\eta_4) + \frac{b+c}{2}(\eta_1\eta_3 + \eta_2\eta_4)$$

$$+ \frac{b-c}{2}(\eta_1\eta_4 + \eta_2\eta_3) \tag{182}$$

$$\Psi_2 = \frac{a+c-i}{2}e^{\vec{\eta}} + \frac{a-c+i}{2}(\eta_1\eta_2 + \eta_3\eta_4) + \frac{b+c+i}{2}(\eta_1\eta_3 + \eta_2\eta_4)$$

$$+ \frac{b-c-i}{2}(\eta_1\eta_4 + \eta_2\eta_3) + \frac{i}{2}(\eta_1 + \eta_4 + \eta_2\eta_3\eta_4 + \eta_1\eta_2\eta_3$$

$$- \eta_2 - \eta_3 - \eta_1\eta_3\eta_4 - \eta_1\eta_2\eta_4) \tag{183}$$

$$\Psi_3 = \frac{a}{2}e^{\eta_1\eta_2+\eta_3\eta_4} + \frac{b+1}{2}(\eta_1\eta_3 + \eta_2\eta_4) + \frac{b-1}{2}(\eta_1\eta_4 + \eta_2\eta_3)$$
$$+ \frac{1}{2}(\eta_3 + \eta_1\eta_2\eta_4 - \eta_4 - \eta_1\eta_2\eta_3) \tag{184}$$

$$\Psi_4 = \frac{a+b}{2}e^{\vec{\eta}} + b(\eta_1\eta_3 + \eta_2\eta_4) + i(-\eta_2\eta_3 + \eta_1\eta_4) + \frac{a-b}{2}(\eta_1\eta_2 + \eta_3\eta_4)$$
$$+ \frac{1}{2}(\eta_2 + \eta_3 + \eta_1\eta_3\eta_4 + \eta_1\eta_2\eta_4 - \eta_1 - \eta_4 - \eta_2\eta_3\eta_4 - \eta_1\eta_2\eta_3) \tag{185}$$

$$\Psi_5 = \frac{a}{2}e^{\eta_1\eta_2+\eta_3\eta_4} - 2i(\eta_2 + \eta_1\eta_4 - \eta_1\eta_2\eta_3) \tag{186}$$

$$\Psi_6 = \frac{a+i}{2}e^{\eta_1\eta_2+\eta_3\eta_4} + \frac{a+i+1}{2}(\eta_1\eta_3 + \eta_2\eta_4) + \frac{a-i-1}{2}(\eta_2\eta_3 + \eta_1\eta_4)$$
$$+ \frac{i+1}{2}(\eta_3 + \eta_1\eta_2\eta_4) + \frac{i-1}{2}(\eta_4 + \eta_1\eta_2\eta_3) - \frac{i}{2}(\eta_1 + \eta_2 + \eta_2\eta_3\eta_4$$
$$+ \eta_1\eta_3\eta_4) \tag{187}$$

$$\Psi_7 = \eta_1\eta_4 + \eta_1\eta_3 + \eta_2\eta_4 - \eta_2\eta_3 + \eta_1\eta_2 + \eta_1\eta_2\eta_3\eta_4 + i(\eta_1\eta_4 + \eta_1\eta_3$$
$$- \eta_2 - \eta_2\eta_3\eta_4 - \eta_1\eta_2\eta_4 + \eta_1\eta_2\eta_3) \tag{188}$$

$$\Psi_8 = \frac{i+1}{2}(e^{\vec{\eta}} - \eta_3 - \eta_1\eta_2\eta_4) + \frac{i-1}{2}(\eta_4 + \eta_1\eta_2\eta_3 - \eta_3\eta_4 - \eta_1\eta_2)$$
$$+ \frac{1}{2}(\eta_2 + \eta_1\eta_4 + \eta_1\eta_3 + \eta_2\eta_3\eta_4 + \eta_1 + \eta_2\eta_3 + \eta_2\eta_4 + \eta_1\eta_3\eta_4)$$
$$- i(\eta_1 + \eta_2\eta_4 + \eta_2\eta_3 + \eta_1\eta_3\eta_4) \tag{189}$$

$$\Psi_9 = \frac{1}{2}e^{\eta_1}(e^{\eta_2\eta_3\eta_4} + \eta_2 + \eta_3\eta_4 + i(\eta_3 + \eta_4 - \eta_2\eta_4 - \eta_2\eta_3)). \tag{190}$$

These states can be represented in terms of trigonometric and exponential η-functions. In such a form their symmetry properties are better pronounced. Namely, we have[24]

$$\Psi_1 = \frac{a+d}{2}e^{\vec{\eta}} + \frac{a-d}{2}(\cos(\eta_1 - \eta_2) - \cos(\eta_3 + \eta_4)) + \frac{b+c}{2}(\cos(\eta_1 - \eta_3)$$
$$- \cos(\eta_2 + \eta_4)) + \frac{b-c}{2}(\cos(\eta_1 - \eta_4) - \cos(\eta_2 + \eta_3)) \tag{191}$$

$$\Psi_2 = \frac{a+c-i}{2}e^{\vec{\eta}} + \frac{a-c+i}{2}(\cos(\eta_1 - \eta_2) - \cos(\eta_3 + \eta_4))$$
$$+ \frac{b+c+i}{2}(\cos(\eta_1 - \eta_3) - \cos(\eta_2 + \eta_4)) + \frac{b-c-i}{2}(\cos(\eta_1 - \eta_4)$$
$$- \cos(\eta_2 + \eta_3)) + \frac{i}{2}(\sin(\eta_1 + \eta_2 + \eta_4) - \sin(\eta_2 + \eta_3 + \eta_4)$$
$$+ \sin(\eta_1 + \eta_3 + \eta_4) - \sin(\eta_1 + \eta_2 + \eta_3)) \tag{192}$$

$$\Psi_3 = \frac{a}{2}e^{\eta_1\eta_2+\eta_3\eta_4} + \frac{b+1}{2}(\cos(\eta_1-\eta_3) - \cos(\eta_2+\eta_4))$$

$$+ \frac{b-1}{2}(\cos(\eta_1-\eta_4) - \cos(\eta_2+\eta_3))$$

$$+ \frac{1}{2}(\sin(\eta_1+\eta_2+\eta_3) - \sin(\eta_1+\eta_2+\eta_4)) \tag{193}$$

$$\Psi_4 = \frac{a+b}{2}e^{\bar\eta} + b(\cos(\eta_1-\eta_3) - \cos(\eta_2+\eta_4)) + i(\cos(\eta_2-\eta_3)$$

$$- \cos(\eta_1+\eta_4) - 2\sin\eta_2\sin\eta_3) + \frac{a-b}{2}(\cos(\eta_1-\eta_2)$$

$$- \cos(\eta_3+\eta_4)) + \frac{1}{2}(\sin(\eta_1-\eta_2-\eta_3+\eta_4) + \sin(\eta_1+\eta_4)$$

$$- \sin(\eta_2+\eta_3)) \tag{194}$$

$$\Psi_5 = \frac{a}{2}e^{\eta_1\eta_2+\eta_3\eta_4} - 2i(\sin\eta_2\cos(\eta_1+\eta_3) + \cos(\eta_1+\eta_4) - 1) \tag{195}$$

$$\Psi_6 = \frac{a+i}{2}e^{\eta_1\eta_2+\eta_3\eta_4} + \frac{a+i+1}{2}(\cos(\eta_1-\eta_3) - \cos(\eta_2+\eta_4)$$

$$+ \frac{a-i-1}{2}(\cos(\eta_2-\eta_3) - \cos(\eta_1+\eta_4)) + \frac{i}{2}(\sin(\eta_1+\eta_3$$

$$+ \eta_4) - \sin(\eta_1+\eta_2+\eta_4) + \sin(\eta_2+\eta_3+\eta_4) - \sin(\eta_1+\eta_2+\eta_3))$$

$$+ \frac{1}{2}(\sin(\eta_1+\eta_2+\eta_3) - \sin(\eta_1+\eta_2+\eta_4)) \tag{196}$$

$$\Psi_7 = \sin\eta_1\sin(\eta_1+\eta_3+\eta_4) + \sin\eta_2\sin(\eta_1-\eta_3+\eta_4)$$

$$+ i(\sin(\eta_2+\eta_3+\eta_4) + \sin(\eta_1-\eta_2-\eta_4) - \sin(\eta_1+\eta_2+\eta_3)) \tag{197}$$

$$\Psi_8 = \frac{1}{2}(\cos(i(\eta_1+\eta_2+\eta_3+\eta_4)) - \sin(\eta_1+\eta_2-\eta_3-\eta_4)$$

$$+ \sin(\eta_1+\eta_2) - \sin(\eta_3+\eta_4) + \sin(\eta_1-\eta_3) + \sin(\eta_2-\eta_4))$$

$$+ \frac{i}{2}(\cos(\eta_1+\eta_2+\eta_3+\eta_4) + \cos(\eta_1+\eta_3+\eta_4)$$

$$- \cos(\eta_2-\eta_3) - 2\sin(\eta_1+\eta_3-\eta_4) + \sin(\eta_1-\eta_2+\eta_3)$$

$$- \sin(\eta_1-\eta_2+\eta_4)) \tag{198}$$

$$\Psi_9 = \frac{1}{2}e^{\eta_1+\eta_2}\cos(\eta_3-\eta_4)) + \frac{i}{2}e^{\eta_1-\eta_2}\sin(\eta_3+\eta_4)) \tag{199}$$

To measure the degree of entanglement of pure states for $n = 4$ one introduces various entanglement measures based on invariants. The basic one is the Cayley determinant $\mathcal{H}$, which as we have already seen has nice form in terms of the η-function. As it was shown[24] the Wronski matrix for $F(\eta_1, \eta_2, \eta_3, \eta_4)$ of the form

$$B_{ij} = \begin{pmatrix} \mathcal{H}(\pi_i \pi_j F) & \tilde{\mathcal{H}}_j(\pi_i F) & \mathcal{H}(\pi_i \partial_j F) \\ \tilde{\mathcal{H}}_i(\pi_j F) & \tilde{\mathcal{H}}_{ij}(F) & \tilde{\mathcal{H}}_i(\partial_j F) \\ \tilde{\mathcal{H}}(\partial_i \pi_j F) & \tilde{\mathcal{H}}_j(\partial_i F) & \mathcal{H}(\partial_i \partial_j F) \end{pmatrix}, \tag{200}$$

where the $\mathcal{H}$, $\tilde{\mathcal{H}}_i$ and $\tilde{\mathcal{H}}_{ij}$ are taken in appropriate form for η-functions with $n = 2, 3, 4$ variables. Taking determinants for matrices B_{12}, B_{13} and B_{14} one gets invariants (new in respect to the previously discussed invariants L, M, N) known from the classical invariant theory and discussed in Ref. 46, which there denoted as $D_{xy}, \ldots$. Namely,

$$D_{xy} = det(B_{12}), \quad D_{xz} = det(B_{13}), \quad D_{xt} = det(B_{14}) \tag{201}$$

The equation (163) yields the relation

$$L^2 + M^2 + N^2 = -2(MN + NL + ML). \tag{202}$$

To get symmetric invariants it is convenient to introduce the following set of invariants

$$W = D_{xy} + D_{xz} + D_{xt}, \tag{203}$$

$$\Sigma = L^2 + M^2 + N^2, \tag{204}$$

$$\Pi = (L - M)(M - N)(N - L). \tag{205}$$

Together with the Cayley determinant they form so called Schläfli basis $\{\mathcal{H}, W, \Sigma, \Pi\}$. For the $n = 4$ states Osterloh and Siewert introduced the entanglement monotones $|\mathcal{F}_i|$; $i = 1, 2, 3, 4, 5$. In terms of the the Schläfli basis they can be written as[48,55]

$$\mathcal{F}_1 = 8(4W - \mathcal{H}^3) \tag{206}$$

$$\mathcal{F}_2 = 16(\mathcal{H}^4 - 4\mathcal{H}W - 4(\mathcal{H}D_{xt} + 4LM)) \tag{207}$$

$$\mathcal{F}_3 = 32(\mathcal{H}^6 - 24\mathcal{H}^2\Sigma - 64\Pi) \tag{208}$$

$$\mathcal{F}_4 = 16(\mathcal{H}^4 - 4\mathcal{H}W - 4(\mathcal{H}D_{xz} + 4LN)) \tag{209}$$

$$\mathcal{F}_5 = 16(\mathcal{H}^4 - 4\mathcal{H}W - 4(\mathcal{H}D_{xy} + 4MN)) \tag{210}$$

The $|\mathcal{F}_i|$ for $i = 2, 4, 5$ separately are not symmetric, but their sum yields the symmetric entanglement monotone

$$|\mathcal{F}_2'| = |\mathcal{F}_2 + \mathcal{F}_4 + \mathcal{F}_5| = 16|3\mathcal{H}^4 - 16\mathcal{H}W + 8\Sigma|, \tag{211}$$

which we prefer as more universal.

Table 1.   Invariants and entanglement monotones $|\mathcal{F}'_2|$, $\mathcal{F}_3$ evaluated on ψ_{GHZ}, ψ_W and ψ_{CW} states.

	ψ_{GHZ}	ψ_W	ψ_{CW}
$|\mathcal{F}'_2|$	3	0	$\frac{11}{9}$
$|\mathcal{F}_3|$	$\frac{1}{2}$	0	$\frac{1}{2}$
$\mathcal{H}$	$\frac{1}{2}$	0	$\frac{1}{2}$
W	0	0	$\frac{1}{72}$
Σ	0	0	0
Π	0	0	0

Let us evaluate $|\mathcal{F}_3|$ and $|\mathcal{F}'_2|$ on the $n = 4$ stats represented by η-function

$$\psi_c(\alpha) \equiv sin\alpha\,\psi_{CW} + cos\alpha\,\psi_{GHZ} \qquad (212)$$

and

$$\psi_s(\alpha) \equiv cos\alpha\,\psi_W + \sin\alpha \star \psi_W. \qquad (213)$$

It is interesting that for $\psi_c(\alpha)$ we get constant value of above entanglement measures while for $\psi_s(\alpha)$ both entanglement monotones vary between zero and its maximal value for the $\psi_s(\alpha)$ family i.e. $|\mathcal{F}_3(\psi_s(\alpha))| = \frac{1}{2}\sin^6(2\alpha))$ and $|\mathcal{F}'_2(\psi_s)| = 3\sin^4(2\alpha)$. Let us look, at the case of the family $\psi_c(\alpha)$. These states belong to the family G_{abcd}. As we know general η-function representing such a state can be written in the form given by Eq. (192). We obtain the $\psi_c(\alpha)$ by taking $a = \cos\alpha + \sin\alpha$, $b = \sin\alpha$, $c = 0$ and $d = \cos\alpha - \sin\alpha$. As it is known two states from the one of the nine families found in Ref. 47 may be in the same orbit, that is why the $|\mathcal{F}_3|$ and $|\mathcal{F}'_2|$ are constant for $\psi_c(\alpha)$. However it is not generic situation. To see this, let us move along G_{abcd} family, taking parameters such that $a = b$ and $c = d$. Denoting

$$\zeta = \left(\frac{2ad}{a^2 + d^2}\right)^2 \qquad (214)$$

we have that

$$|\mathcal{F}_3(\psi_{ad})| = |\frac{1}{2} - \frac{3}{2}\zeta^4 + \zeta^6| = |(\zeta - 1)(\zeta + \frac{1}{2})| \qquad (215)$$

and

$$|\mathcal{F}'_2(\psi_{ad})| = |3 - 2^2\zeta + \zeta^2| = |(\zeta - 1)(\zeta - 3)| \qquad (216)$$

Again, when one of the parameters a or d vanishes, we obtain invariant $|\mathcal{F}_3(\psi_{ad})|$ and $|\mathcal{F}_2'(\psi_{ad})|$, but otherwise these entanglement monotones vary for states from the G_{abcd}. In particular we find states for which above entanglement measures simultaneously vanish (taking into account (214) we see that both polynomials have only one, common zero). Hence the entanglement monotones $|\mathcal{F}_3|$, $|\mathcal{F}_2'|$ indicate that the states

$$\psi_a = a(e^{\vec{\eta}} + (\cos(\eta_1 - \eta_3) - \cos(\eta_2 + \eta_4))) \tag{217}$$

and

$$\psi_d = d(\cos(\eta_1 - \eta_2) - \cos(\eta_3 + \eta_4) + \cos(\eta_1 - \eta_4) - \cos(\eta_2 + \eta_3)) \tag{218}$$

do not exhibit genuine four-qubit entanglement. Indeed, $\psi_a \sim \psi_W^{(13)} \psi_W^{(24)}$ and $\psi_d \sim \psi_{GHZ}^{(13)} \psi_{GHZ}^{(24)}$. So, there is only residual two-qubit entanglement.

The question of the proper genuine monotones for the system of $n = 4$ qubits is still open, but as present analysis shows considered above monotones due to their symmetry properties and behavior on η-trigonometric states there is an indication that they are a reasonable candidates.

7. Conclusions

In these notes we make extended exposition of the nilpotent commuting variables approach to quantum mechanics which provides new tool suitable for studying the entanglement questions of qubit systems. Qubit has two dimensional representations and commuting nilpotent variables automatically provide correct properties of the single qubit and multiqubit systems. Of course one can consider a higher order nilpotency, then in the context of quantum mechanics we arrive to description of qutrits etc. (cf. contribution of Mandilara and Akulin to the present volume).

On the other hand, there is developed formalism suitable to describe classical systems with nilpotent commuting coordinates,[23] in analogy to he psedo-mechanical systems described by anticommuting variables. The peculiarities of appropriate differential calculus[23,25] and an analog of the variational calculus were studied.[23] Hence, by now, various building blocks of the theory involving nilpotent commuting variables are at hand.

We think that this formalism it is something more then an ad hoc tool, and given here the η-toolbox might be useful.

Acknowledgements

This work is partially supported by the Wrocław University grant 2349/W/IFT/10.

Bibliography

1. A. M. Frydryszak, Int. J. Mod. Phys. **A 25**, 951–983 (2010).
2. K. Ziegler, *Europhys. Lett.* **9** (1989) 277.
3. K. Ziegler, *Physica A* **179** (1991) 301.
4. K. Ziegler, *Phys. Lett. B* **46** (1992) 6647.
5. K. Ziegler, *J. Stat. Phys.* **64** (1991) 277.
6. K. Ziegler, *"Phase Transitions of a Bose Gas in an Optical Lattice"* arXiv:cond-mat/0303593v1.
7. K. Ziegler, *J. Phys.: Condens. Matter* **17** (2005) S1809.
8. F. Palumbo, Phys. Rev. **D 50**, R1917–R1920 (1993).
9. F. Palumbo, Phys. Rev. **D 50**, 2826–2829 (1994).
10. F. Palumbo, Nucl. Phys. **B 37** (Proc. Suppl.), 522–524 (1994).
11. S. Caracciolo and F. Palumbo, *Nucl. Phys. B (Proc. Suppl.)* **63A-C** (1998) 790–792.
12. M. B. Barbaro, A. Molinari, and M. R. Quaglia, Phys. Rev. C **64**, 011302(R) (2001).
13. G. De Franceschi and F. Palumbo, *Mod. Phys. Lett.* **A 11** (1995) 901.
14. M. B. Barbaro, A. Molinari, F. Palumbo, and M. R. Quaglia *Phys. Lett.* **B 476** (2000) 477–487.
15. M. B. Barbaro, A. Molinari, and M. R. Quaglia *Phys. Rev.* **C 64** (2001) 011302(R).
16. S. Caracciolo, V. Laliena, and F. Palumbo *JHEP 02* **64** (2007) 034.
17. M. B. Barbaro, R. Cenni, S. Chiacchiera, A. Molinari, and F. Palumbo, *Ann. Phys.* **322** (2007) 2665.
18. A. Mandiliara, V. M. Akulin, A. V. Smilga, and L. Viola, Phys. Rev. A **74**, 022331 (2006).
19. M. A. Nielsen and I. L. Chuang, *Quantum Computation and Information* (Cambridge University Press, Cambridge, England, 2000).
20. H. S. Green, Phys. Rev. **90**, 270–273 (1953).
21. Y. Ohnuki and S. Kamefuchi, *Quantum Field Theory and Parastatistics* (University of Tokyo Press, Tokyo 1982).
22. L-A. Wu and D. A. Lidar, J. Math. Phys. **43**, 4506–4525 (2002).
23. A. M. Frydryszak, Int. J. Mod. Phys. **A 22**, 2513–2533 (2007).
24. A. M. Frydryszak, *"Nilpotent quantum mechanics, qubits and flavors of entanglement"*, arXiv:0810.3016 [quant-ph].
25. A. T. Filippov, A. P. Isaev and A. B. Kurdikov: *Theor. Math. Phys.* **94**, 150–165 (1993).
26. A. M. Frydryszak, Czech. J. Phys. **56**, 1155–1161 (2006).
27. A. M. Frydryszak, Rep. Math. Phys. **61**, 227–235 (2008).
28. A. Mandiliara and L. Viola, J. Phys. B: At. Mol. Opt. Phys. **40** S167–S180 (2007).
29. S. B. Bravyi and A. Yu. Kitaev *Ann. Phys* **298** 210 (2000).
30. M. Čadek and J. Šimša, Aequationes Math. **40**, 8–25 (1990); M. Čadek and J. Šimša, Czech. Math. J. **41**, 342–358 (1991).
31. G. S. Staples, Advances Appl. Clifford Alg. **15**, 213–232 (2005); G. S. Staples and R. Schott, *Clifford algebras and random graphs* Prepublication de

l'Institut Élie Caratan **37**, (2005); G. S. Staples and R. Schott, Eur. J. Comb. doi:10.1016/j.ejc.2007.07.003; G. S. Staples and R. Schott, J. Theor. Probab. **20**, 257–274 (2007).

32. N. Wallach *"Quantum computing and entanglement for matematicians"* (unpublished lecture notes).
33. D. Meyer and N. Wallach, *"Invariants for multiple qubits: the case of 3 qubits"* (Mathematics of computation, Comput. Math. Ser., Chapman & Hall/CRC, Boca Raton, FL, 77–97, 2002).
34. G. Mahler G and V. A. Weberruß, *Quantum Networks (Dynamics of Open Nanostructures)* (Springer, Berlin, 1995).
35. J.-Y. Choi and S.-I. Hong, Phys. Rev. A **60**, 796–799 (1999).
36. D. A. Meyer and N. R. Wallach, J. Math. Phys. **43**, 4273 (2002).
37. C. Emary, J. Phys. A: Math. Gen. **37**, 8293–8302 (2004).
38. L. Amico, R. Fazio, A. Osterloh, and V. Vedral, Rev. Mod. Phys. **80**, 517–576 (2008).
39. K. Eckert, J. Schliemann, and D. Brußand M. Lewenstein, Ann. Phys (N.Y.) **299**, 88–127 (2002).
40. S. Cyparissos, Rend. Circ. Mat. Palermo **18**, 360–362 (1906).
41. T. M. Rassias, Bull. Inst. Math. Acad. Sinica **14**, 377–382 (1986).
42. F. Neuman, Linear Algebra Appl. **134**, 153–164 (1990).
43. A. Prástaro and T. M. Rassias, J. Comp. Appl. Math. **113**, 93–122 (2000).
44. A. Miyake, Phys. Rev. A **67**, 012108 (2003); **67**, 012108 (2003); A. Miyake, F. Verstraete, Phys. Rev. A **69**, 012101 (2004).
45. A. Miyake, and M. Wadati, Quantum Inf. Comput. **2**, 540–546 (2002).
46. J. G. Luque and J. Y. Thibon, Phys. Rev. A **67**, 042303 (2003).
47. F. Verstraete, J. Dehaene, B. DeMoor, and H. Verschelde, Phys. Rev. A **65**, 052112 (2002).
48. A. Osterloh and J. Siewert, *"Constructing N-qubit entanglement monotones from antilinear operators"*, `quant-ph/0410102`.
49. P. Levay, J. Phys. A: Math and General **37**, 1821–1841 (2004).
50. W. K. Wootters, Phys. Rev. Lett. **80**, 2245 (1998).
51. V. Coffman, J. Kundu, and W. K. Wooters, Phys. Rev A **61**, 052306 (2000).
52. G. Brennen, QIC **3**, 619 (2003), `quant-ph/0305094`.
53. O. Chterental and D. Z. Doković, *Linear Algebra Reasearch Advances* (Nova Science, Hauppauge, 2007) (`quant-ph/0612184`).
54. F. Verstraete, J. Dehaene, and B. De Moor, Phys. Rev. A **68**, 012103 (2003).
55. X.-J. Ren, W. Jiang, X. Zhou, Z.-W. Zhou, and G.-C. Guo, *"Permutation invariant monotones for multipartite entanglement characterization"*, `arXiv:0804.3461v1 [quant-phys]`.

Experiments on quantum coherence with cold atoms

Wojciech Gawlik* and Adam Wojciechowski

*M. Smoluchowski Institute of Physics, Jagiellonian University,
Reymonta 4, PL-30-059 Kraków, Poland*
E-mail: gawlik@uj.edu.pl

Coherence, or superposition of atomic states gives rise to a plethora of interesting interference effects in the interaction of light and atoms. It drastically modifies optical properties of atomic media and finds applications in quantum information and metrology. This lecture provides an introduction to the coherence effects in atomic samples, presents some experiments on quantum coherences with cold atoms and discusses their possible applications.

Keywords: quantum coherence, cold atoms.

1. Introduction

Introduction of tunable lasers resulted in a rapid development of atomic physics. Use of these modern light sources providing intense, narrowband, coherent and tunable light enabled observation of many new phenomena. With off-the-shelf products one can now study various atomic processes and reach temperatures much lower than anywhere in the Universe. Ultra cold atomic samples enable not only high-resolution spectroscopy but also studies of new phenomena in quantum degenerate gases.

The term *coherence*, commonly associated with light, also has well established meaning of quantum superpositions of atomic states and is usually divided into the optical coherences (associated with the optical transitions) and Zeeman coherences (between the magnetic sublevels). The latter might be created in the meta-stable (ground) states, offering very long lifetimes.

The aim of this lecture is to introduce the coherence effects in atomic samples, present some experiments on quantum coherences with cold atoms and discuss their possible applications.

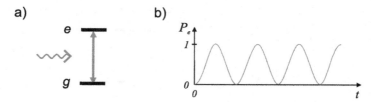

Figure 1.   (a) Two-level atom coupled to the resonant light. (b) Probability of finding the atom in the excited state as a function of time. The periodic probability oscillations are the Rabi oscillations (spontaneous emission is neglected here).

2. What is coherence?

2.1. *Two-level atom*

The simplest system in which atomic coherence can be observed is a two-level atom (TLA) with ground state $|g\rangle$ and excited state $|e\rangle$, schematically shown in Fig. 1(a). The state of such a system at any time can be expressed in the $\{|e\rangle, |g\rangle\}$ basis as

$$|\psi(t)\rangle = g(t)\,|g\rangle + e(t)\,|e\rangle, \tag{1}$$

where $g(t)$ and $e(t)$ are the time-dependent coefficients, with normalization $|g|^2 + |e|^2 = 1$. The system's dynamics can be calculated using the Schrödinger equation and the initial conditions.

Suppose the atom initially in the ground state, i.e., $|\psi\rangle = |g\rangle$ starts to interact with the light wave. The solution of this problem is a periodic *Rabi* oscillation. Figure 1(b) shows the coherent dynamics of the system, i.e., the probability $P_e = |g(t)|^2$ of finding the atom in the excited state. For a resonant light, $P_e(t)$ is reaching the value of 1 and reduces its amplitude for the detuned light.

Alternative, and more general, description of the state of the system can be given in terms of density matrix. The density matrix formalism becomes particularly useful when dealing with the atomic ensembles where the number of Schrödinger equations to solve could be very large. The state of the TLA system in this formalism is described by

$$\rho = \begin{pmatrix} \rho_{gg} & \rho_{ge} \\ \rho_{eg} & \rho_{ee} \end{pmatrix}, \tag{2}$$

where $\rho = |\psi\rangle\langle\psi|$, while the dynamics are governed by the *master equation*

$$\dot{\rho} = -\frac{i}{\hbar}\,[H, \rho], \tag{3}$$

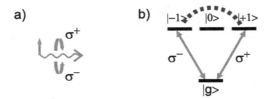

Figure 2.   Illustration of the two-level atom with angular momentum $J_e = 1$ in the excited state and $J_g = 0$ in the ground state coupled to the linearly polarized resonant light. The linear σ polarization is a coherent superposition of the σ_+ and σ_- circular polarizations that drive the $|g\rangle \rightarrow |+1\rangle$ and $|g\rangle \rightarrow |-1\rangle$ transitions, respectively. The dashed line represents the Zeeman coherence that is created by the light.

where H is the Hamiltonian and the brackets represent the commutator. The normalization requires that the trace of this matrix must be equal to 1 and the matrix is hermitian, i.e. $\rho_{eg} = \rho_{ge}^*$. Diagonal elements of such a matrix represent the probability of finding the atom in a given state and thus are called populations of the states. On the other hand, the off-diagonal elements are the measure of the co-evolution of the two states and are called *coherences*. Since Eq. (1) is the generic form of a superposition state, the off-diagonal elements of the density matrix, the coherences, are associated with the superposition states.

2.2. *Two-level atom with angular momentum*

In this subsection we will extend the TLA model by assuming $J_e=1$ value of the excited state angular momentum and $J_g = 0$ in the ground state. To investigate the interaction of light with such a system one has to define a quantization axis and take into account appropriate selection rules. Suppose the light is linearly polarized and we choose a quantization axis along its propagation direction, as shown in Fig. 2. In this case the light is σ polarized and it is useful to treat this polarization as a coherent superposition of the two circular polarizations, σ^+ and σ^-, each coupling to its own transition. The $|0\rangle$ state is not coupled and will be neglected in the following. The density matrix of such a system has a form:

$$\rho = \begin{pmatrix} \rho_{gg} & \rho_{g-} & \rho_{g+} \\ \rho_{-g} & \rho_{--} & \rho_{-+} \\ \rho_{+g} & \rho_{+-} & \rho_{++} \end{pmatrix}. \tag{4}$$

Without going into calculations,[1] the immediate result would be a non-zero value of the optical coherences ρ_{g-} and ρ_{g+} as in Sec. 2.1. However, the

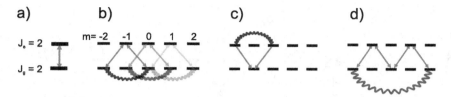

Figure 3.   Illustration of the two-level atom with non-zero angular momenta in both ground and excited states (a) and possible coherences arrangements: independent Λ systems (b), *V*-like system (c), double Λ with $\Delta m = 4$ coherence (d).

solution of the master equation leads also to a non-zero value of the ρ_{-+} coherence, which is a result of the simultaneous and coherent driving of the two optical transitions. The emerging coherence between two magnetic sublevels is named the *Zeeman coherence* and corresponds to the

$$|\Phi(t)\rangle = a(t)\,|-1\rangle + b(t)\,|+1\rangle \tag{5}$$

superposition state. It has a characteristic time dependence which is an important experimental signature in the form of

$$\rho_{-+} \propto exp\left[-\frac{i}{\hbar}\left(E_{+1} - E_{-1}\right)t\right], \tag{6}$$

where $E_{\pm 1}$ is the energy of the $|\pm 1\rangle$ state.

The single, linearly polarized photon can create Zeeman coherence between sublevels differing by 2 in their magnetic number ($\Delta m = 2$). However, one can now extend this model to higher angular momentum structures, both in the ground and excited states, as well as multi-photon interactions. Figure 3 shows the possible scenarios for the $J_g, J_e = 2$ atom. Firstly, the independent Λ-like systems can be formed leading to the three Zeeman coherences in the ground state. Analogously, the *V*-like structures lead to the existence of Zeeman coherences in the excited state. Finally, the $\Delta m = 4$ coherence may arise as a multi-photon process, in which the coherence is generated firstly between pairs of Zeeman sublevels with $\Delta m = 2$ and then subsequent interactions merge them into a new one.

3. Manifestation of the coherence

The existence of coherence in a given quantum system may lead to an important modification of the optical properties of the macroscopic medium. The common examples are the slow (or stopped) light experiments,[2] electromagnetically induced transparency,[3] etc. In this section several types of related experiments are discussed.

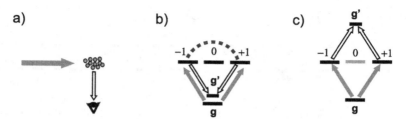

Figure 4.   Illustration of the light scattering experiment in which atom is excited by the linearly polarized light and laterally emitted fluorescent light is detected. (a) the arrangement, (b) schematics of the atomic structure with excitation (solid red arrows) and fluorescence (empty black arrows) paths and Zeeman coherence between the $|\pm1>$ sublevels (broken blue line), (c) the same paths, shown to resemble Young's double slit experiment.

3.1. *Light scattering*

Perhaps the simplest experiment in which the Zeeman coherence can be observed is the scattering of the light shown in Fig. 4 which is based on measuring intensity of resonance fluorescence vs. the magnetic field intensity (the Hanle effect[4,5]). The linearly polarized light creates a coherence in the excited state of the $J_g = 0$, $J_e = 1$ atom (broken line in Fig. 4). The intensity I of the fluorescence light is then given by

$$I \propto \sum_{g'} \left| \langle g' | \hat{D} | \Phi \rangle \right|^2, \tag{7}$$

where $\hat{D}$ stands for the dipole moment operator, $|\Phi\rangle$ is given by Eq. (5) and the sum goes over all possible final states. In our case, the only possible final state is $|g\rangle$, and the intensity becomes

$$I \propto \underbrace{\rho_{--}|D_{g-1}|^2}_{I_1} + \underbrace{\rho_{++}|D_{g+1}|^2}_{I_2} + \underbrace{\rho_{-+}D_{g-1}D_{g+1}^* + \rho_{+-}D_{g+1}D_{g-1}^*}_{I_{12}}, \tag{8}$$

where I_{12} is the interference term which represents the coherence contribution to the scattered light intensity. It can be shown that the Zeeman coherence has an amplitude given by

$$\rho_{-+} \propto \frac{e^{-i\gamma t}}{E_{+1} - E_{-1} - i\hbar\gamma} \exp\left[-\frac{i}{\hbar}(E_{+1} - E_{-1})t \right], \tag{9}$$

where γ is the coherence decay rate (the inverse of the lifetime τ). Thus, by changing the energies of the $|\pm1\rangle$ levels, e.g. by a magnetic field, it is possible to vary the contribution of the interference term to the scattered

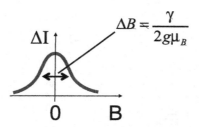

Figure 5. Intensity of the scattered light as a function of the magnetic field.

light. In the case of degenerate levels, the coherence remains stationary and has the maximal (yet decaying) amplitude

$$\rho_{-+} \propto \frac{e^{-i\gamma t}}{-i\hbar\gamma}. \tag{10}$$

If the Zeeman sublevels degeneracy is lifted by the applied magnetic field B, the coherence is oscillating. In the experiment where continuous wave light beam is interacting with an atomic ensemble, it is necessary to perform averaging over time, as the coherence in individual atoms is created at random times and evolves during lifetime τ. As a result, the fluorescence intensity varies by ΔI which has a form of a Lorentzian resonance centered at $B = 0$ when plotted as a function of the magnetic field (Fig. 5).

3.2. *Electromagnetically induced transparency*

One of the most popular coherence effects is known as the *electromagnetically induced transparency* (EIT).[3] The idea behind the effect is such that sufficiently strong and resonant light may induce a spectral transparency window for the other light beam. The effect is most often observed in a three-level cascade system with a strong light field resonantly coupling the upper atomic states and a weak light probing absorption on the transition between the lower states. Figure 6 (a) shows an exemplary three-level system in which another kind of EIT can be observed, associated with the, so called, *coherent population trapping*, CPT.[6] In this case, two beams of the opposite helicity are coupling the $m = 1$ and $m = -1$ magnetic sublevels with the single excited state. If the beams are sufficiently strong, the observed absorption and dispersion spectra for either of the beams are strongly modified, as seen in Fig. 6 (b).

The modification of the absorption and dispersion profiles results from existence of the coherence between the two ground states introduced by the presence of both laser beams and has far-reaching consequences. Rapid

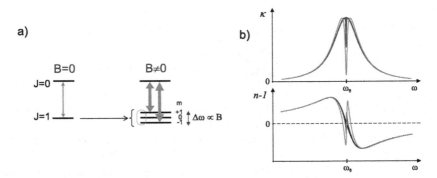

Figure 6.   (a) Schematics of the energy levels structure without (left) and with the magnetic field (right). (b) Absorption coefficient κ and refractive index n spectra without (blue) and with (red) the coherence of the ground-state sublevels.

change of the refractive index with the light frequency results in modification of the group velocity of light that passes through the medium. EIT experiments showed group velocities as low as 8 m/s in a thermal sample of rubidium vapor.[7]

3.3. Nonlinear Faraday Effect

Linear Faraday effect (LFE) is a well-known magneto-optical phenomenon, in which the light polarization plane rotates when propagating through a medium in a longitudinal magnetic field. The difference of the refractive indices $n_\pm$ for the σ_+ and σ_- circular polarizations results in different phases acquired by the two helicities over the sample length l and, as a consequence, in the rotation of the linear polarization plane

$$\theta = (n_+ - n_-)l\frac{2\pi}{\lambda}, \tag{11}$$

where λ is the light wavelength. The dependence of θ on the magnetic field is hidden here in the difference of the refractive indices. For a given wavelength and a medium without resonant transitions this can be expressed in a form

$$\theta = VlB, \tag{12}$$

where B is the magnetic field strength and V is the material (Verdet) constant. On the other hand, for the near-resonant light, Eq. (11) yields a familiar Faraday angle dependence on the magnetic field

$$\theta(B) \propto \frac{\Delta\omega(B)}{\Delta\omega^2(B) + \gamma^2/4}, \tag{13}$$

where γ is the transition linewidth and $\Delta\omega$ is the shift of the transition frequency for one of the circular polarizations. For not-too-strong magnetic fields the latter is linear in the magnetic field and thus the rotation amplitude has a form of a dispersive Lorentzian with the width corresponding to the optical transition width, Doppler-broadened in the case of gas samples. The asymmetric $\theta(B)$ dependence is called the Faraday resonance. In the central part, the resonance curve can be approximated with a straight line the slope of which has a meaning of the Verdet constant. It is particularly important for magnetometric applications. The slope increases with the increasing amplitude and decreasing width of the $\theta(B)$ curve yielding better accuracy of the magnetic field measurement. Typically, the resonant atomic gas samples have Verdet constants a few orders of magnitude higher than non-resonant solid-state samples, even though typical atomic densities are only on the order of $10^9 - 10^{10}$ at/cm^3.

Application of strong, resonant light might additionally result in the creation of coherence between Zeeman sublevels of the ground state.[1,8] Their presence manifests in the magneto-optical rotation resonances similar to that in LFE, however, having much narrower widths due to the longer lifetimes of the atomic ground states (Fig. 7).

For resonant excitation, rotation angle θ is a measure of circular birefringence, $\theta \propto (n_+ - n_-)$, where $n_\pm$ are the refractive indices for $\sigma^\pm$ polarized light and

$$n_\pm - 1 \propto \mathcal{E}^{-1} \sum_{eg} Re(d_{eg}^{(\pm)} \rho_{eg}^{(\pm)}) \tag{14}$$

with $\mathcal{E}$ being the light electric field amplitude, $d_{eg}^\pm$ the matrix element of the dipole moment associated with the $\sigma^\pm$-polarized light-beam components, and $\rho_{eg}^\pm$ the related density matrix elements. The summation goes over all ground- and excited-state sublevels g and e linked by the allowed transitions. In the stationary regime, ρ_{eg} can be expressed as

$$\rho_{eg}^{(\pm)} = \frac{1}{\delta_{eg} - i\Gamma/2} \sum_{e'g'} \left(\Omega_{eg'}^{(\pm)} \rho_{g'g} - \rho_{ee'} \Omega_{e'g}^{(\pm)} \right), \tag{15}$$

where $\delta_{\alpha\beta}$ and $\Omega_{\alpha\beta}$ denote respectively the light detuning and Rabi frequency for the $\alpha \leftrightarrow \beta$ transition, and $\Gamma/2$ is the relaxation rate of the optical coherence. The polarization index $\pm$ is related with magnetic quantum numbers of states e, g by standard selection rules for polarized light. Relations (14, 15) indicate that optical coherences, and consequently also

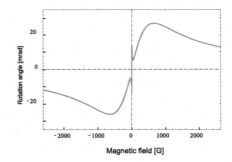

Figure 7.   Rotation angle versus magnetic field. The central, narrow structure is the nonlinear Faraday rotation resonance.

the refractive indices and rotation angle, depend on the density matrix elements $\rho_{g'g}$ and $\rho_{ee'}$ which represent populations of and coherences between Zeeman sublevels of the ground and excited states. For not-too-strong light, the excited-state coherences are negligible and the rotation signal becomes sensitive mainly to the ground-state coherences.

Main decoherence processes for the atomic vapor samples are the atomic collisions or collisions with the vapor cell walls. Diluted atomic vapor in a buffer-gas or a paraffin-coated vapor cells can reach coherence life-times as long as a fraction of a second, that results in μG-wide resonances, which can be used for ultra-sensitive measurements of the magnetic field. Current state-of-art optical magnetometers have a sensitivity comparable to that of *superconducting quantum interference devices* (SQUIDs) and the Verdet constant reaches 10^{10} rad/T·m.[1,9]

4. Nonlinear Faraday Effect with cold atoms

The nonlinear Faraday effect (NFE) has been widely studied in alkali-metal atomic vapor cells.[1] However, the use of cold atomic samples at temperatures of tens of μK offers some advantages over the room-temperature experiments. Firstly, the negligibility of the Doppler broadening allows for precise addressing of a single optical transition. Moreover, the cold samples offer low relaxation rates and high spatial resolution as they are typically confined in a small volume of the trap. Finally, depending on the trapping procedure, a range of optical depths of the sample can be achieved (typically 1-100). On the other hand, cold-atom experiments have also their disadvantages. Not only they are more technically challenging (laser cooling, trap switching, etc.) but are also more prone to various systematic errors, such

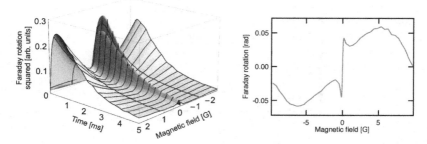

Figure 8. Nonlinear Faraday rotation in a cold-atom sample recorded at 2 ms of probing time. The wide structure is the linear- and the narrow central resonance is the nonlinear Faraday effect. Pictures from Ref. 10.

as caused by light pressure. In the following, we present the results obtained with the cold-atom sample of ^{85}Rb released from the magneto-optical trap (MOT).[10]

The experimental sequence, in which NFE signals were recorded consisted of three phases. Firstly, the sample was captured and cooled in a standard MOT configuration. Subsequently, the cooling (trapping) laser, as well as the MOT quadrupole magnetic field were switched off. In addition, during the preparation phase, the magnetic field of a given value was applied along the probing beam direction (Faraday field). Finally, the probing beam was switched on and its polarization plane rotation was measured. After typically 10 ms, the probing beam and Faraday field were switched off and the trap fields were switched on to recapture the expanding atomic cloud. The sequence was repeated for each Faraday field value.

Typical resonance observed in the experiment is shown in Fig. 8. It consists of a wide structure of about 10 G width associated with a linear Faraday effect and a central, narrow feature which is a signature of the nonlinear Faraday rotation. The central resonance of the 20 mG width corresponds to the presence of $\Delta m = 2$ coherences in the $F = 3$ ground state of ^{85}Rb.

By using a special time sequence of the light beams and Faraday magnetic field, the time evolution of the superposition state could be studied. Figure 9 depicts typical time dependences obtained for two different light intensities at B corresponding to the maximum coherence contribution, compared with the time dependence of a linear Faraday rotation at 3 G. While the linear signal rises almost instantaneously (within the time resolution of our detector), the NFE signal, i.e. the quantum coherence, needs

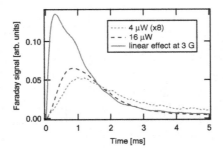

Figure 9.   Time dependence of the NFE resonances recorded for two laser powers: 4μW (magnified 8 times) and 16μW at a magnetic field of 45 mG compared to the LFE rotation at 3 G. Picture from Ref. 10.

some time to develop. As seen, the time becomes shorter for higher light intensity. Such experiment allows diagnostics of superposition states created in cold-atom sample and can be useful for applications in quantum information and in precision magnetometry. The narrow width of the feature (some mG with the present setup) shows the potential of NFE with cold atoms for precision magnetometry with important prospective features: μG sensitivity, large dynamic range (zero-field to several G), and sub-mm spatial resolution in magnetic field mapping.

4.1. *High-field magnetometry*

The stationary ground-state coherences are destroyed when Larmor precession becomes faster than the coherence relaxation time, hence direct observation of the NFR signals is limited to a very narrow range around $B = 0$. For observation of NFE not only around the zero magnetic field it is necessary to use amplitude modulation of light.[11] In this arrangement, strobed pumping creates the modulated Zeeman coherence and phase sensitive detection is used to extract the magneto-optical rotation amplitude. In addition to the zero-field resonance, two other resonances appear in the demodulated rotation signal when the modulation frequency Ω_m matches $\pm$ twice Larmor precession frequency in a given magnetic field. These high-field resonances result from the optical pumping synchronous with the Larmor precession. The width of these resonances is determined by the coherence lifetime and, in case of long-lived ground states, can be as narrow as the zero field resonance. Figure 10 shows NFE signal with two AMOR resonances at ± 3 G that are the evidence of driving $|\Delta m| = 2$ coherences at non-zero magnetic fields. Presence of such high-field resonances allows

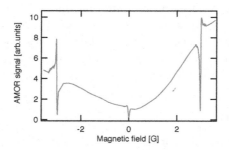

Figure 10.   Nonlinear Faraday rotation with amplitude modulated light (AMOR) in a cold-atom sample. The plot shows the square of the rotation angle. The side resonances occur when the modulation frequency equals twice the Larmor frequency in a given magnetic field. Picture from Ref. 10.

for precision magnetometry of non-zero magnetic fields.

5. Quantum state engineering and quantum degeneracy

Various techniques of quantum state engineering, i.e. controlling the quantum-state superpositions, have been developed for hot atomic vapor samples. They include creation and selective detection of coherences with specific Δm including higher-order coherences with $\Delta m > 2$,[12,13] competition between NFE and EIT,[14] and superposition of various coherences.[15] Most of these techniques should also be useful in experiments with cold atoms. One important requirement is the reduction of all factors limiting the coherence lifetime. A promising solution is the application of optical dipole traps for such experiments.

An exciting emerging field of research on quantum coherence is experiments in the quantum-degeneracy regime, e.g. with the Bose-Einstein condensate (BEC) in which superposition states of matter-waves can be studied. EIT has been directly measured in a pump-probe experiment in a BEC by Ahufinger et al.[16] It has been also applied by L. Hau and coworkers for spectacular "slow and halted light" experiments[17] and demonstration of quantum memory with about 1 sec storing time.[18]

6. Conclusions

The atomic superposition states (coherences) can be created and manipulated with a high degree of control and reproducibility. Since such states are pivotal for quantum information, the related experiments pave the way to realization of many ideas of quantum information. On the other hand, there are many practical applications of coherently prepared media. One specific field is quantum magnetometry. Cold atoms allow for extending the coherence lifetimes and make it possible to study the phenomena with quantum-degenerate matter waves.

Acknowledgments

This work is partially supported by the Polish Ministry of Science and Education (grant # N N202 046337, and within the program supporting the National Laboratory of AMO Physics in Toruń, Poland) and in part by the Foundation for Polish Science (Team Programme).

Bibliography

1. W. Gawlik and S. Pustelny, Nonlinear Faraday effect and its applications, in *New trends in quantum coherence and nonlinear optics*, ed. R. Drampyan, Horizons in World Physics, Vol. 263 (Nova Science Publishers, New York, 2009) pp. 45–82.
2. Z. Dutton, N. S. Ginsberg, C. Slowe and L. V. Hau, *Europhysics News* **35**, p. 33 (2004).
3. M. Fleischhauer, A. Imamoglu and J. Marangos, *Reviews of Modern Physics* **77**, 633 (2005).
4. W. Hanle, *Z. Physik* **30**, 93 (1924).
5. G. Moruzzi and F. Strumia (eds.), *The Hanle Effect And Level-Crossing Spectroscopy* (Plenum Press, New York, 1991).
6. E. Arimondo, in *Progress in Optics*, ed. E. Wolf (Elsevier, Amsterdam, 1996) pp. 259–354.
7. D. Budker, D. F. Kimball, S. M. Rochester and V. V. Yashchuk, *Physical Review Letters* **83**, 1767 (1999).
8. D. Budker, W. Gawlik, D. F. Kimball, S. M. Rochester, V. V. Yashchuk and A. Weis, *Reviews of Modern Physics* **74**, 1153 (2002).
9. D. Budker and M. Romalis, *Nature Physics* **3**, 227 (2007).
10. A. Wojciechowski, E. Corsini, J. Zachorowski and W. Gawlik, *Physical Review A* **81**, 053420 (2010).
11. W. Gawlik, L. Krzemień, S. Pustelny, D. Sangla, J. Zachorowski, M. Graf, A. O. Sushkov and D. Budker, *Applied Physics Letters* **88**, 131108 (2006).
12. V. V. Yashchuk, D. Budker, W. Gawlik, D. F. Kimball, Y. P. Malakyan and S. M. Rochester, *Physical Review Letters* **90**, 253001 (2003).
13. S. Pustelny, W. Gawlik, S. M. Rochester, D. F. J. Kimball, V. V. Yashchuk and D. Budker, *Physical Review A* **74**, 063420 (2006).
14. R. Drampyan, S. Pustelny and W. Gawlik, *Physical Review A* **80**, p. 033815 (2009).
15. V. M. Acosta, M. Auzinsh, W. Gawlik, P. Grisins, J. M. Higbie, D. F. J.

Kimball, L. Krzemien, M. P. Ledbetter, S. Pustelny, S. M. Rochester, V. V. Yashchuk and D. Budker, *Optics Express* **16**, 11423 (2008).

16. V. Ahufinger, R. Corbalan, F. Cataliotti, S. Burger, F. Minardi and C. Fort, *Optics Communications* **211**, 159 (2002).

17. L. V. Hau, S. E. Harris, Z. Dutton and C. H. Behroozi, *Nature* **397**, 594 (1999).

18. R. Zhang, S. R. Garner and L. V. Hau, *Physical Review Letters* **103**, 233602 (2009).

Dynamical entanglement of three-level atoms

Lech Jakóbczyk

Institute of Theoretical Physics
University of Wrocław
Plac Maxa Borna 9, 50-204 Wrocław, Poland
E-mail: ljak@ift.uni.wroc.pl

The dynamical creation of entanglement between three-level atoms coupled to the common vacuum is investigated. We show that in the case of closely separated atoms collective damping can generate robust entanglement of the asymptotic states. For a large class of initial states the asymptotic entanglement is distillable.

Keywords: three-level atoms; entanglement production; distillable entanglement.

1. Introduction

Dynamical creation of entanglement by the indirect interaction between otherwise decoupled systems has been studied by many researchers mainly in the case of two-level atoms interacting with the common vacuum. When the two atoms are separated by a distance small compared to the radiation wave length λ, there is a substantial probability that a photon emitted by one atom will be absorbed by the other and the resulting process of photon exchange produces correlations between the atoms. Such correlations may cause that initially separable states become entangled (see e.g. Refs. 1–4). The case of three-level atoms is very interesting for many reasons. First of all, in a system of coupled multi-level atoms having closely lying energy states and interacting with the vacuum, quantum interference between different radiative transitions can occur, resulting in coherences in a system which are known as *vacuum-induced coherences*. Beside the usual effects such as collective damping and dipole-dipole interaction involving non-orthogonal transition dipole moments, [5,6] radiative coupling can produce here a new interference effect in the spontaneous emission. This effect manifests by the cross coupling between radiative transitions with *orthogo-*

nal dipole moments [7] and is strongly dependent on the relative orientation of the atoms. [8,9] All such collective properties of the system influence the quantum dynamics, which can significantly differ from a corresponding single atom dynamics. On the other hand, the theory of entanglement between the pairs of such atoms is much more complex then in the case of qubits. As is well known, there is no simple necessary and sufficient condition of entanglement, since the Peres-Horodecki separability criterion [10,11] only shows that the states which are not positive after partial transposition (NPPT states) are entangled. But there can exist entangled states which are positive after this operation and such states are not distillable. [12] Hence all entangled states can be divided into two classes, one contains free entangled states that can be distilled using local operations and classical communication (LOCC) and the other consist of bound entangled states for which no LOCC strategy can be used to extract pure state entanglement.

In this paper we review the results concerning the dynamics of entanglement between three-level atoms with vacuum induced coherences, obtained in Refs. 13 and 14. We have shown there that for small distance between the atoms the system decays to a stationary state which can be entangled, even if the initial state was separable. [13] On the other hand, if the distance is comparable to the radiation wavelength, the dynamics brings all initial states into the asymptotic state in which both atoms are in their ground states but still there can be some transient entanglement between the atoms. [14]

In the present paper, we focus on the process of dynamical creation of distillable entanglement due to the collective damping and cross coupling between the three-level atoms. For the specific initial states and small interatomic distance we show that the asymptotic states are both entangled and distillable. It happens also for some bound entangled initial state. Thus we show that the physical process of spontaneous emission can transform initial bound entanglement into free distillable entanglement of the asymptotic state. For larger distances, the dynamics of this initial state is very peculiar: the system very quickly disentangle and only after some finite time there suddenly appears a distillable entanglement. (The similar phenomenon of delayed sudden birth of entanglement was observed in the case of two-level atoms. [16]) So also in this situation the creation of transient distillable entanglement occurs.

2. Mixed-state entanglement and distillation

2.1. *Distillability of entanglement*

Distillability of mixed entangled state ρ is the property that enables to convert n copies of ρ into less number of k copies of maximally entangled pure state by means of LOCC.[17] It is known that all pure entangled states can be reversibly distilled [18] and any mixed two-qubit entangled state is also distillable. [19] In general case, the following necessary and sufficient condition for entanglement distillation was shown in Ref. 12 : the state ρ is distillable if and only if there exists n such that ρ is n-copy distillable i.e. $\rho^{\otimes n}$ can be filtered to a two-qubit entangled state. This condition is however hard to apply, since conclusions based on a few copies may be misleading. [20] More practical but not necessary condition is based on the reduction criterion of separability. [21] The criterion can be stated as follows: if a bipartite state ρ of a compound system AB is separable, then

$$\rho_A \otimes \mathbb{1} - \rho \geq 0 \quad \text{and} \quad \mathbb{1} \otimes \rho_B - \rho \geq 0 \tag{1}$$

where

$$\rho_A = \mathrm{tr}_B\, \rho, \quad \rho_B = \mathrm{tr}_A\, \rho$$

As was shown in Ref. 22, any state that violates (1) is distillable, so if

$$\rho_A \otimes \mathbb{1} - \rho \not\geq 0 \quad \text{or} \quad \mathbb{1} \otimes \rho_B - \rho \not\geq 0 \tag{2}$$

the state ρ can be distilled. The condition (2) is easy to check and we will use it in our discussion of dynamical aspects of distillability.

2.2. *Peres-Horodecki criterion and bound entanglement*

To detect entangled states of two qutrits, we apply Peres-Horodecki criterion of separability. [10,11] From this criterion follows that any state ρ for which its partial transposition ρ^{PT} is non-positive (NPPT state), is entangled. One defines also *negativity* of the state ρ as

$$N(\rho) = \frac{||\rho^{\mathrm{PT}}||_{\mathrm{tr}} - 1}{2} \tag{3}$$

$N(\rho)$ is equal to the absolute value of the sum of the negative eigenvalues of ρ^{PT} and is an entanglement monotone,[23] however it cannot detect entangled states which are positive under partial transposition (PPT states). Such states exist [24] and as was shown in Ref. 12, are not distillable. They are called *bound entangled PPT states*. Up to now, it is not known if there exist bound entangled NPPT states.[25]

To detect some of bound entangled PPT states we can use the realignment criterion of separability. [26,27] The criterion states that for any separable state ρ of a compound system, the matrix $R(\rho)$ with elements

$$\langle m| \otimes \langle \mu|R(\rho)|n\rangle \otimes |\nu\rangle = \langle m| \otimes \langle n| \rho |\mu\rangle \otimes |\nu\rangle \tag{4}$$

has a trace norm not greater then 1. So if the *realignment negativity* defined by

$$N_R(\rho) = \max\left(0, \frac{||R(\rho)||_{\text{tr}} - 1}{2}\right) \tag{5}$$

is greater then zero, the state ρ is entangled. In the case of two qubits, the measure (5) cannot detect all NPPT states, [28] but for larger dimension the criterion is capable of detecting some bound entangled PPT states. [26]

3. Time evolution of three-level atoms

To study the dynamics of entanglement between three-level atoms we consider the model introduced by Agarwal and Patnaik. [7] We start with the short description of the model. Consider two identical three-level atoms (A and B) in the V configuration. The atoms have two near-degenerate excited states $|1_\alpha\rangle$, $|2_\alpha\rangle$ ($\alpha = A, B$) and ground states $|3_\alpha\rangle$. Assume that the atoms interact with the common vacuum and that transition dipole moments of atom A are parallel to the transition dipole moments of atom B. Due to this interaction, the process of spontaneous emission from two excited levels to the ground state take place in each individual atom but a direct transition between excited levels is not possible. Moreover, the coupling between two atoms can be produced by the exchange of the photons, but in such atomic system there is also possible the radiative process in which atom A in the excited state $|1_A\rangle$ loses its excitation which in turn excites atom B to the state $|2_B\rangle$. This effect manifests by the cross coupling between radiation transitions with orthogonal dipole moments. The evolution of this atomic system can be described by the following master equation [7]

$$\frac{d\rho}{dt} = (L^A + L^B + L^{AB})\rho \tag{6}$$

where for $\alpha = A, B$ we have

$$L^\alpha\rho = \sum_{k=1}^{2} \gamma_{k3}\left(2\sigma_{3k\alpha}\rho\sigma_{k3\alpha} - \sigma_{a3\alpha}\sigma_{3k\alpha}\rho - \rho\sigma_{k3\alpha}\sigma_{3k\alpha}\right) \tag{7}$$

and

$$
L^{AB}\rho = \sum_{k=1}^{2} \sum_{\alpha=A,B} \Gamma_{k3} \left(2\sigma_{3k}^{\alpha}\rho\sigma_{k3}^{\neg\alpha} - \sigma_{k3}^{\neg\alpha}\sigma_{3k}^{\alpha}\rho - \rho\sigma_{k3}^{\neg\alpha}\sigma_{3k}^{\alpha} \right)
$$

$$
+ i \sum_{k=1}^{2} \Omega_{k3} \left[\sigma_{k3}^{A}\sigma_{3k}^{B} + \sigma_{k3}^{B}\sigma_{3k}^{A}, \rho \right]
$$

$$
+ \Gamma_{vc} \sum_{\alpha=A,B} \left(2\sigma_{31}^{\alpha}\rho\sigma_{23}^{\neg\alpha} - \sigma_{23}^{\neg\alpha}\sigma_{31}^{\alpha}\rho - \rho\sigma_{23}^{\neg\alpha}\sigma_{31}^{\alpha} \right. \tag{8}
$$

$$
\left. + 2\sigma_{32}^{\alpha}\rho\sigma_{13}^{\neg\alpha} - \sigma_{13}^{\neg\alpha}\sigma_{32}^{\alpha}\rho - \rho\sigma_{13}^{\neg\alpha}\sigma_{32}^{\alpha} \right)
$$

$$
+ i\,\Omega_{vc} \sum_{\alpha=A,B} \left[\sigma_{23}^{\alpha}\sigma_{31}^{\neg\alpha} + \sigma_{32}^{\alpha}\sigma_{13}^{\neg\alpha}, \rho \right]
$$

In the equations (7) and (8), $\neg\alpha$ is A for $\alpha = B$ and B for $\alpha = A$, σ_{jk}^{α} is the transition operator from $|k_\alpha\rangle$ to $|j_\alpha\rangle$ and the coefficient γ_{j3} represents the single atom spontaneous-decay rate from the state $|j\rangle$ ($j = 1, 2$) to the state $|3\rangle$. Since the states $|1_\alpha\rangle$ and $|2_\alpha\rangle$ are closely lying, the transition frequencies ω_{13} and ω_{23} satisfy

$$
\omega_{13} \approx \omega_{23} = \omega_0
$$

Similarly, the spontaneous-decay rates

$$
\gamma_{13} \approx \gamma_{23} = \gamma
$$

The coefficients Γ_{j3} and Ω_{j3} are related to the coupling between two atoms and are the collective damping and the dipole-dipole interaction potential, respectively. The coherence terms Γ_{vc} and Ω_{vc} are cross coupling coefficients, which couple a pair of orthogonal dipoles. Detailed analysis shows the cross coupling between two atoms strongly depend on the relative orientation of the atoms and there are such configurations of the atomic system that $\Gamma_{vc} = \Omega_{vc} = 0$ and the other configurations for which $\Gamma_{vc} \neq 0$, $\Omega_{vc} \neq 0$. Moreover, all the coupling coefficients are small for large distance R between the atoms and tend to zero for $R \to \infty$. On the other hand, when $R \to 0$, Ω_{13}, Ω_{23} and Ω_{vc} diverge, whereas

$$
\Gamma_{13}, \Gamma_{23} \to \gamma, \quad \text{and} \quad \Gamma_{vc} \to 0
$$

The time evolution of the initial state of the system is given by the semi-group $\{T_t\}_{t\geq 0}$ of completely positive mappings acting on density matrices, generated by $L^A + L^B + L^{AB}$. The properties of the semi-group crucially depends on the distance R between the atoms and the geometry of the system. Irrespective to the geometry, when R is large compared to the radiation wavelength, the semi-group $\{T_t\}_{t\geq 0}$ is uniquely relaxing with the asymptotic state $|3_A\rangle \otimes |3_B\rangle$. Thus, for any initial state its entanglement asymptotically approaches 0. But still there can be some transient entanglement between the atoms. On the other hand, in the strong correlation regime (when $R \to 0$), the semi-group is not uniquely relaxing and the asymptotic stationary states are non-trivial and depend on initial conditions. The explicit form of the asymptotic state ρ_{as} for any initial state ρ with matrix elements ρ_{kl} (with respect to the canonical basis) was found in Ref. 13. It is given by

$$
\rho_{as} = \begin{pmatrix}
0 & 0 & 0 & 0 & 0 & 0 & 0 & 0 & 0 \\
0 & 0 & 0 & 0 & 0 & 0 & 0 & 0 & 0 \\
0 & 0 & x & 0 & 0 & z & -x & -z & w \\
0 & 0 & 0 & 0 & 0 & 0 & 0 & 0 & 0 \\
0 & 0 & 0 & 0 & 0 & 0 & 0 & 0 & 0 \\
0 & 0 & \overline{z} & 0 & 0 & y & -\overline{z} & -y & v \\
0 & 0 & -x & 0 & 0 & -z & x & z & -w \\
0 & 0 & -\overline{z} & 0 & 0 & -y & \overline{z} & y & -v \\
0 & 0 & \overline{w} & 0 & 0 & \overline{v} & -\overline{w} & -\overline{v} & t
\end{pmatrix}
\tag{9}
$$

where

$$
x = \frac{1}{8}\left(\rho_{22} + 2\rho_{33} + \rho_{44} + 2\rho_{77} - 2\operatorname{Re}\rho_{24} - 4\operatorname{Re}\rho_{37}\right)
$$

$$
z = \frac{1}{4}\left(\rho_{36} - \rho_{38} - \rho_{76} + \rho_{78}\right)
$$

$$
w = \frac{1}{4}\left(\rho_{26} + \rho_{28} + 2\rho_{39} - \rho_{46} - \rho_{48} - 2\rho_{79}\right)
\tag{10}
$$

$$
y = \frac{1}{8}\left(\rho_{22} + \rho_{44} + 2\rho_{66} + 2\rho_{88} - 2\operatorname{Re}\rho_{24} - 4\operatorname{Re}\rho_{68}\right)
$$

$$
v = \frac{1}{4}\left(-\rho_{23} - \rho_{27} + \rho_{43} + \rho_{47} + 2\rho_{69} - 2\rho_{89}\right)
$$

and

$$
t = 1 - 2x - 2y
$$

To get some insight into the process of creation of non-trivial asymptotic state ρ_{as}, it may be useful to consider the basis of collective states in $\mathbb{C}^9$, given by the doubly excited states

$$
|e_1\rangle = |1_A\rangle \otimes |1_B\rangle, \quad |e_2\rangle = |2_A\rangle \otimes |2_B\rangle
$$

the ground state

$$
|g\rangle = |3_A\rangle \otimes |3_B\rangle
$$

and generalized symmetric and antisymmetric Dicke states

$$|s_{kl}\rangle = \frac{1}{\sqrt{2}} \left[|k_A\rangle \otimes |l_B\rangle + |l_A\rangle \otimes |k_B\rangle \right]$$

$$|a_{kl}\rangle = \frac{1}{\sqrt{2}} \left[|k_A\rangle \otimes |l_B\rangle - |l_A\rangle \otimes |k_B\rangle \right] \tag{11}$$

where $k, l = 1, 2, 3\,;\, k < l$. The states (11) are entangled, but in contrast to the case of two-level atoms, they are not maximally entangled. One can also check that the doubly excited states $|e_1\rangle$, $|e_2\rangle$ and the symmetric Dicke states $|s_{kl}\rangle$ decay to the ground state $|g\rangle$, whereas antisymmetric states $|a_{13}\rangle$ and $|a_{23}\rangle$ decouple from the environment and therefore are stable. Moreover, the state $|a_{12}\rangle$ is not stable, but it is asymptotically non-trivial. Notice that the collective states can be used to the direct characterization of the asymptotic behaviour of the system. In particular, the parameters x and y in (9) are given by the populations in the antisymmetric states $|a_{13}\rangle$, $|a_{23}\rangle$ and $|a_{12}\rangle$:

$$x = \frac{1}{4} \left(\langle a_{12}|\rho|a_{12}\rangle + 2 \langle a_{13}|\rho|a_{13}\rangle \right)$$

$$y = \frac{1}{4} \left(\langle a_{12}|\rho|a_{12}\rangle + 2 \langle a_{23}|\rho|a_{23}\rangle \right)$$

The remaining parameters can be computed in terms of the coherences between the collective states. Since the populations $\langle a_{13}|\rho|a_{13}\rangle$ and $\langle a_{23}|\rho|a_{23}\rangle$ are stationary, the states which have the property of trapping the initial populations in $|a_{13}\rangle$ or $|a_{23}\rangle$, create the non-trivial asymptotic state ρ_{as} with the stationary entanglement. On the other hand, the population in the state $|a_{12}\rangle$ is not stable, but it can be transformed into $\langle a_{13}|\rho|a_{13}\rangle$ and $\langle a_{23}|\rho|a_{23}\rangle$ in such a way that the values of the parameters x and y are fixed.

4. Generation of stationary distillable entanglement

As was shown in Ref. 13, the negativity of the asymptotic states (9) can be obtained analytically in the case of diagonal (i.e. separable) initial states. For such states, only the parameters x, y and t are non-zero and the asymptotic negativity reads

$$N(\rho_{\mathrm{as}}) = \frac{1}{2} \left[\sqrt{4(x^2 + y^2) + t^2} - t \right] \tag{12}$$

Note that every nontrivial asymptotic states from that class is entangled. Consider now the explicit examples of such behaviour of the system. Let the initial state be a pure separable state of the form

$$|\Psi\rangle = |j_A\rangle \otimes |k_B\rangle, \quad j, k = 1, 2, 3 \tag{13}$$

It is obvious that the states $|1_A\rangle \otimes |1_B\rangle$ and $|2_A\rangle \otimes |2_B\rangle$ decay to the ground state $|g\rangle$. On the other hand, the initial state $|1_A\rangle \otimes |3_B\rangle$ (atom A in the excited state and atom B in the ground state) has the population in the Dicke state $|a_{13}\rangle$ which is equal to $\frac{1}{2}$, thus for that state

$$x = \frac{1}{4}, \quad t = \frac{1}{2} \quad \text{and} \quad y = z = w = v = 0$$

and the asymptotic state is entangled with negativity

$$N(\rho_{\text{as}}) = \frac{\sqrt{2} - 1}{4} \tag{14}$$

Similarly, the state $|2_A\rangle \otimes |3_B\rangle$ has the population $\frac{1}{2}$ in the state $|a_{23}\rangle$ and also produces asymptotic state with the same value of entanglement. The same behaviour can be observed for the initial states $|3_A\rangle \otimes |1_B\rangle$ and $|3_A\rangle \otimes |2_B\rangle$.

When two atoms are initially in different excited states i.e. we have the states $|1_A\rangle \otimes |2_B\rangle$ or $|2_A\rangle \otimes |1_B\rangle$, then the initial populations in the states $|a_{13}\rangle$ and $|a_{23}\rangle$ are equal to zero, but the population in the non-stable state $|a_{12}\rangle$ is non-zero and equals $\frac{1}{2}$. During the evolution this population is transformed into the states $|a_{13}\rangle$ and $|a_{23}\rangle$, in such a way that the asymptotic state ρ_{as} satisfies

$$\langle a_{12}|\rho_{\text{as}}|a_{12}\rangle = 0$$

and

$$\langle a_{13}|\rho_{\text{as}}|a_{13}\rangle = \langle a_{23}|\rho_{\text{as}}|a_{23}\rangle = \frac{1}{4}$$

Thus the state ρ_{as} is also entangled, but its negativity is less then (14) and equals $(\sqrt{6} - 2)/8$.

Now we prove that all asymptotic states corresponding to the diagonal initial states are distillable. To do this, we show that such ρ_{as} violate reduction criterion (1). Indeed, since for these states

$$
\mathrm{tr}_B \rho_\mathrm{as} \otimes \mathbb{1} - \rho_\mathrm{as} =
\begin{pmatrix}
x & 0 & 0 & 0 & 0 & 0 & 0 & 0 & 0 \\
0 & x & 0 & 0 & 0 & 0 & 0 & 0 & 0 \\
0 & 0 & 0 & 0 & 0 & 0 & x & 0 & 0 \\
0 & 0 & 0 & y & 0 & 0 & 0 & 0 & 0 \\
0 & 0 & 0 & 0 & y & 0 & 0 & 0 & 0 \\
0 & 0 & 0 & 0 & 0 & 0 & 0 & y & 0 \\
0 & 0 & x & 0 & 0 & 0 & a & 0 & 0 \\
0 & 0 & 0 & 0 & 0 & y & 0 & b & 0 \\
0 & 0 & 0 & 0 & 0 & 0 & 0 & 0 & c
\end{pmatrix}
\tag{15}
$$

where

$$
a = 1 - 2x - y, \quad b = 1 - x - 2y, \quad c = x + y
$$

and the matrix on the right hand side of (15) has two negative leading principal minors (other minors are positive), so

$$
\mathrm{tr}_B \, \rho_\mathrm{as} \otimes \mathbb{1} - \rho_\mathrm{as} \not\geq 0
$$

Similarly

$$
\mathbb{1} \otimes \mathrm{tr}_A \, \rho_\mathrm{as} - \rho_\mathrm{as} \not\geq 0
$$

The interesting examples of nontrivial asymptotic states are given by the separable initial states where the one atom is in the excited state and the other is in the ground state or two atoms are in different excited states. In all such cases, the created entanglement is free and can be distilled.

Now we consider the possibility of creating free stationary entanglement from the bound initial entanglement. As the initial states we take the family introduced in Ref. 29

$$
\rho_\alpha = \frac{2}{7} \, |\Psi_0\rangle\langle\Psi_0| + \frac{\alpha}{7} \, P_+ + \frac{5 - \alpha}{7} \, P_-, \quad 3 < \alpha \leq 4
\tag{16}
$$

where

$$
|\Psi_0\rangle = \frac{1}{\sqrt{3}} \sum_{j=1}^{3} |j_A\rangle \otimes |j_B\rangle,
$$

$$
P_+ = \frac{1}{3} \left(P_{|1_A\rangle\otimes|2_B\rangle} + P_{|2_A\rangle\otimes|3_B\rangle} + P_{|3_A\rangle\otimes|1_B\rangle} \right)
$$

and

$$
P_- = \frac{1}{3} \left(P_{|2_A\rangle\otimes|1_B\rangle} + P_{|3_A\rangle\otimes|2_B\rangle} + P_{|1_A\rangle\otimes|3_B\rangle} \right).
$$

The states (16) have positive partial transposition but are entangled, as can be shown by computing the realignment negativity. For ρ_α it is given by

$$N_R(\rho_\alpha) = \frac{1}{21} \left(\sqrt{3\alpha^2 - 15\alpha + 19} - 1 \right) \tag{17}$$

and is obviously positive for $3 < \alpha \leq 4$.

Although the states (16) are not diagonal, one can check that the corresponding asymptotic states have the same form as in the diagonal case. In fact, for all initial states ρ_α there is only one asymptotic state ρ_{as} given by

$$x = y = \frac{5}{56} \quad \text{and} \quad t = \frac{9}{14}$$

By the above discussion, this state is entangled and moreover its entanglement is distillable.

5. Delayed creation of distillable entanglement

In this section we study in details the evolution of entanglement of the bound entangled initial states (16) for $3 < \alpha \leq 4$, beyond the strong correlation regime. In that case, the asymptotic state is trivial, but some transient entanglement between the atoms can be produced. For simplicity, we consider such atomic configuration for which the cross coupling coefficients are equal to zero. One can check that the initial states (16) will evolve into the states of the form

$$\rho_\alpha(t) = \begin{pmatrix} \rho_{11} & 0 & 0 & 0 & \rho_{15} & 0 & 0 & 0 & \rho_{19} \\ 0 & \rho_{22} & 0 & 0 & 0 & 0 & 0 & 0 & 0 \\ 0 & 0 & \rho_{33} & 0 & 0 & 0 & \rho_{37} & 0 & 0 \\ 0 & 0 & 0 & \rho_{44} & 0 & 0 & 0 & 0 & 0 \\ \rho_{51} & 0 & 0 & 0 & \rho_{55} & 0 & 0 & 0 & \rho_{59} \\ 0 & 0 & 0 & 0 & 0 & \rho_{66} & 0 & \rho_{68} & 0 \\ 0 & 0 & \rho_{73} & 0 & 0 & 0 & \rho_{77} & 0 & 0 \\ 0 & 0 & 0 & 0 & 0 & \rho_{86} & 0 & \rho_{88} & 0 \\ \rho_{91} & 0 & 0 & 0 & \rho_{95} & 0 & 0 & 0 & \rho_{99} \end{pmatrix} \tag{18}$$

where all non-zero matrix elements are time dependent.

Numerical analysis indicates that during the time evolution the realignment negativity (5) of the initial state very rapidly goes to zero, so the system almost immediately disentangle. To consider possible creation of free entanglement, let us first check if PPT condition can be violated during such evolution. After taking the partial transposition, the state (18) becomes

$$
\rho_\alpha(t)^{\mathrm{PT}} = \begin{pmatrix}
\rho_{11} & 0 & 0 & 0 & 0 & 0 & 0 & 0 & \rho_{37} \\
0 & \rho_{22} & 0 & \rho_{15} & 0 & 0 & 0 & 0 & 0 \\
0 & 0 & \rho_{33} & 0 & 0 & 0 & \rho_{19} & 0 & 0 \\
0 & \rho_{51} & 0 & \rho_{44} & 0 & 0 & 0 & 0 & 0 \\
0 & 0 & 0 & 0 & \rho_{55} & 0 & 0 & 0 & \rho_{68} \\
0 & 0 & 0 & 0 & 0 & \rho_{66} & 0 & \rho_{59} & 0 \\
0 & 0 & \rho_{91} & 0 & 0 & 0 & \rho_{77} & 0 & 0 \\
0 & 0 & 0 & 0 & 0 & \rho_{95} & 0 & \rho_{88} & 0 \\
\rho_{37} & 0 & 0 & 0 & \rho_{86} & 0 & 0 & 0 & \rho_{99}
\end{pmatrix} \tag{19}
$$

One can check that determinant d of the matrix (19) equals

$$
\begin{aligned}
d = &(\rho_{22}\rho_{44} - |\rho_{15}|^2)(\rho_{33}\rho_{77} - |\rho_{19}|^2)(\rho_{66}\rho_{88} - |\rho_{59}|^2) \\
&\times (\rho_{11}\rho_{55}\rho_{99} - \rho_{55}|\rho_{37}|^2 - \rho_{11}|\rho_{86}|^2)
\end{aligned} \tag{20}
$$

We can show numerically that (20) changes the sign, since the last factor

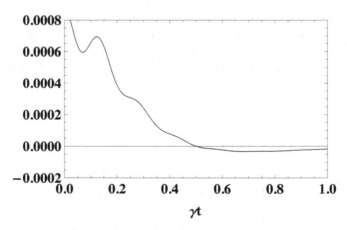

Figure 1.   The time evolution of the last factor in (20) for the initial state (16) with $\alpha = 3.6$ and the interatomic distance $R = 0.2\,\lambda$.

is positive for all $t < t_N$ and becomes negative if $t > t_N$, for some $t_N > 0$, and the remaining factors are positive. Moreover, all other leading principal minors of the matrix (19) are always positive. So the evolution of the bound entangled state (16) has the interesting property: for all $t < t_N$ the states $\rho_\alpha(t)$ are PPT and then suddenly they become NPPT states (see Fig. 1).

Now we discuss distillability of the states $\rho_\alpha(t)$. Since we cannot exclude the possibility that there are NPPT states which are non-distillable, we try to apply the reduction criterion of entanglement. As we know from the discussion in Sect. 2, any state violating this criterion is necessarily distillable. By direct computations we show that the matrix

$$\text{tr}_B\, \rho_\alpha(t) \otimes \mathbb{1} - \rho_\alpha(t)$$

equals to

$$
\begin{pmatrix}
r_{11} & 0 & 0 & 0 & -\rho_{15} & 0 & 0 & 0 & -\rho_{19} \\
0 & r_{22} & 0 & 0 & 0 & 0 & 0 & 0 & 0 \\
0 & 0 & r_{33} & 0 & 0 & 0 & -\rho_{37} & 0 & 0 \\
0 & 0 & 0 & r_{44} & 0 & 0 & 0 & 0 & 0 \\
-\rho_{15} & 0 & 0 & 0 & r_{55} & 0 & 0 & 0 & -\rho_{59} \\
0 & 0 & 0 & 0 & 0 & r_{66} & 0 & -\rho_{68} & 0 \\
0 & 0 & -\rho_{73} & 0 & 0 & 0 & r_{77} & 0 & 0 \\
0 & 0 & 0 & 0 & 0 & -\rho_{86} & 0 & r_{88} & 0 \\
-\rho_{91} & 0 & 0 & 0 & -\rho_{95} & 0 & 0 & 0 & r_{99}
\end{pmatrix}
\tag{21}
$$

where

$$
r_{kk} =
\begin{cases}
\rho_{11} + \rho_{22} + \rho_{33} - \rho_{kk}, & k = 1, 2, 3 \\
\rho_{44} + \rho_{55} + \rho_{66} - \rho_{kk}, & k = 4, 5, 6 \\
\rho_{77} + \rho_{88} + \rho_{99} - \rho_{kk}, & k = 7, 8, 9
\end{cases}
\tag{22}
$$

We compute that leading principal minors of the matrix (21) which can change the sign during the evolution, and find the following expressions

$$
\begin{aligned}
m_5 &= r_{22}r_{33}r_{44}\left(r_{11}r_{55} - |\rho_{15}|^2\right) \\
m_6 &= r_{22}r_{33}r_{44}r_{66}\left(r_{11}r_{55} - |\rho_{15}|^2\right) \\
m_7 &= r_{22}r_{44}r_{66}\left(r_{11}r_{55} - |\rho_{15}|^2\right)\left(r_{33}r_{77} - |\rho_{37}|^2\right) \\
m_8 &= r_{22}r_{44}\left(r_{11}r_{55} - |\rho_{15}|^2\right)\left(r_{33}r_{77} - |\rho_{37}|^2\right)\left(r_{66}r_{88} - |\rho_{68}|^2\right)
\end{aligned}
\tag{23}
$$

where m_k for $k = 5, 6, 7, 8$ are determinants of principal $k \times k$ submatrices of the matrix (21). It turns out that the factor $r_{11}r_{55} - |\rho_{15}|^2$ is always positive, but as follows from the numerical analysis, $r_{33}r_{77} - |\rho_{37}|^2$ as well as $r_{66}r_{88} - |\rho_{68}|^2$ change the sign during the evolution (see Fig. 2). Let t_D be the time at which the factor $r_{66}r_{88} - |\rho_{68}|^2$ changes the sign. We see that $t_D > 0$, so only after that time the matrix (21) becomes non-positive. It means that for $t > t_D$, the states $\rho_\alpha(t)$ are necessarily distillable. One can

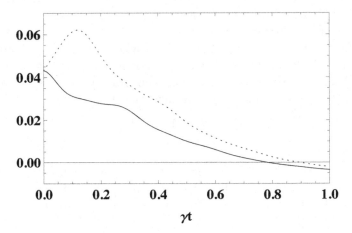

Figure 2.   The time evolution of $r_{33}r_{77} - |\rho_{37}|^2$ (dotted line) and $r_{66}r_{88} - |\rho_{68}|^2$ (solid line) for the initial state (16) with $\alpha = 3.6$ and $R = 0.2\,\lambda$.

check that $t_D > t_N$, and we see that the initial bound entangled state (16) evolves in the remarkable way: for all $t \leq t_N$ it is PPT, for $t_N < t \leq t_D$ it is NPPT but a priori can be non-distillable and only after t_D it becomes distillable. To show this in the explicit way, let us introduce the measure of the violation of reduction criterion, defined as

$$N_{\mathrm{red}}(\rho) = \max\left(0,\, -\lambda_{\mathrm{min}}^{\mathrm{red}}\right) \qquad (24)$$

where $\lambda_{\mathrm{min}}^{\mathrm{red}}$ is the minimal eigenvalues of the matrix

$$\rho_{\mathrm{red}} = \mathrm{tr}_B\, \rho \otimes \mathbb{1} - \rho$$

The quantity (24) can be called the reduction negativity of the state ρ. For the bound entangled initial state (16) the evolution of negativity and reduction negativity is given below (Fig. 3). So we observe in the system the phenomenon of delayed sudden birth of distillable entanglement. The numerical value of t_D depends on the choice of the parameter α and the interatomic distance R. For the initial state with $\alpha = 3.6$ and the distance $R = 0.2\,\lambda$, we obtain $t_D\gamma \approx 0.78$ whereas $t_N\gamma \approx 0.49$.

6. Conclusions

We have studied the dynamics of entanglement in the system of three-level atoms in the V configuration, coupled to the common vacuum. In the case

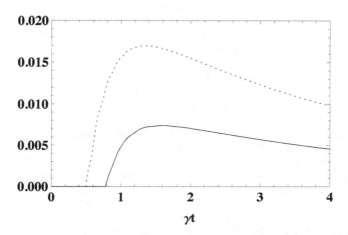

Figure 3. The time evolution negativity N (dotted line) and reduction negativity N_{red} (solid line) for the initial state (16) with $\alpha = 3.6$ and $R = 0.2\,\lambda$.

of small (compared to the radiation wavelength) separation between the atoms, the system has nontrivial asymptotic states which can be entangled even if the initial states were separable. For the large class of separable initial states the asymptotic states are not only entangled but also distillable. The same is true for some class of bound entangled initial states. Thus we have shown that the dynamics of the system can transform bound entanglement into the free distillable entanglement of stationary states. For the atoms separated by larger distances only some transient entanglement can exist but still the dynamical generation of entanglement is possible. We have shown that this happens also for the class of bound entangled initial states. Moreover we have demonstrated that such states evolve in a very peculiar way: they almost immediately disentangle after the atoms begin to interact with the vacuum, then for some finite period of time there is no entanglement and suddenly at some time the entanglement starts to build up. But this entanglement a priori can be nondistillable. We have analysed this problem using the reduction criterion of separability and found that the free entanglement surely appears in the system after some additional period of time.

Bibliography

1. L. Jakóbczyk, J. Phys. A **35**, 6383(2002); **36**, 1537(2003), Corrigendum.
2. Z. Ficek, R. Tanaś, J. Mod. Opt. **50**, 2765(2003).

3. R. Tanaś, Z. Ficek, J. Opt. B, **6**, S90(2004).
4. L. Jakóbczyk, J. Jamróz, Phys. Lett. A **318**, 318(2003).
5. G. S. Agarwal, *Quantum Statistical Theories of Spontaneous Emission and Their Relation to Other Approaches* (Springer, Berlin, 1974).
6. Z. Ficek and S. Swain, *Quantum Interference and Coherence: Theory and Experiments* (Springer, Berlin, 2005).
7. G. S. Agarwal, A. K. Patnaik, Phys. Rev. A **63**, 043805(2001).
8. J. Evers, M. Kiffner, M. Macovei and Ch. H. Keitel, Phys. Rev. A **73**, 023804(2006).
9. S.I. Schmid and J. Evers, Phys. Rev. A **77**, 013822(2008).
10. A. Peres, Phys. Rev. Lett. **77**, 1413(1996).
11. M. Horodecki, P. Horodecki, R. Horodecki, Phys. Lett. A **223**, 1(1996).
12. M. Horodecki, P. Horodecki, R. Horodecki, Phys. Rev. Lett. **80**, 5239(1998).
13. Ł. Derkacz, L. Jakóbczyk, J. Phys. A **41**, 205304(2008).
14. Ł. Derkacz, L. Jakóbczyk, Phys. Lett. A **372**, 7117(2008).
15. Z. Sun, X.G. Wang, Y.B. Gao, C.P. Sun, Eur. Phys. J. D**46**, 521(2008).
16. Z. Ficek, R. Tanaś, Phys. Rev. A**77**, 054301(2008).
17. C.H. Bennett, G. Brassard, S. Popescu, B. Schumacher, J.A. Smolin, W.K. Wootters, Phys. Rev. Lett. **76**, 722(1996).
18. C.H. Bennett, H.J. Bernstein, S. Popescu, B. Schumacher, Phys. Rev. A **53**, 2046(1996).
19. M. Horodecki, P. Horodecki, R. Horodecki,Phys. Rev. Lett. **78**, 574(1997).
20. J. Watrous, Phys. Rev. Lett. **93**, 010502(2004).
21. N.J. Cerf, C. Adami, M.R. Gingrich, Phys. Rev. A **60**, 898(1999).
22. M. Horodecki, P. Horodecki, Phys. Rev. A **59**, 4026(1999).
23. G. Vidal, R.F. Werner, Phys. Rev. A **65**, 032314(2002).
24. P. Horodecki, Phys. Lett. A **232**, 333(1997).
25. R.Horodecki, P. Horodecki, M. Horodecki, K. Horodecki, Rev. Mod. Phys. **81**, 865(2009).
26. K. Chen, L.-A. Wu, Quantum Inf. Comput. **3**, 193(2003).
27. O. Rudolph, J. Phys. A **36**, 5825(2003).
28. O. Rudolph, Phys. Rev. A **67**, 032312(2003).
29. P. Horodecki, M. Horodecki, R. Horodecki, Phys. Rev. Lett. **82**, 1056(1999).

PPT states and measures of entanglement

Władysław A. Majewski

Institute of Theoretical Physics and Astrophysics
Gdańsk University
Wita Stwosza 57
80-952 Gdańsk, Poland
E-mail: fizwam@univ.gda.pl

Entanglement is one of the most important concepts in Quantum Information. But, mathematically one of the most hard task in quantum theory is to describe entanglement. Although the progress in this field is remarkable, still the theory is not complete. In this lecture we describe some aspects of the theory of positive maps, then the full characterization of PPT states will be given. Finally, we show how far we are able to classify quantum entanglement.

Keywords: entanglement; positive maps; PPT states; quantum correlations.

1. Introduction

Characterization of quantum entanglement, so also PPT states (states having positive partial transposition) and measures of entanglement, is closely related with the structure of the set of positive maps (see [14], [24]). Moreover, it is also related with a classification of states of a composite system (see [1], and [15]). The main difficulty in carrying out these tasks is that the characterization of the structure of positive maps is not complete (see [30], [24], [22], and [23]).

For simplicity of these notes, which are based on the joint work with T. Matsuka and M. Ohya (see [25]), we restrict ourselves only to the composite systems consisting of two parties, i.e. to bipartite systems. The aim of this lecture is to give an exposition of a characterization of PPT states and measures of entanglements. To this end, we very briefly outline, most essential for our purposes, aspects of the theory of positive maps. In this part we also establish our notation and provide the necessary vocabulary. Then, in Section 3 we will look more closely at PPT-states. We summarize without proofs the full and operational characterization of PPT states, for the proofs we refer the reader to [25]. Section 4 is devoted to the quantification of entanglement, i.e. to a characterization of some useful measures of entanglement. It is worth pointing out that the characterization of measurements of entanglements in terms of Hilbert spaces offers a great simplification (see Section 4).

However, before going to discuss these matters we must describe some preliminary notions regarding quantum systems as well as quantum maps.

2. Positive maps

In this section we compile some basic facts on the theory of positive maps on the ordered structures and an emphasis will be put on algebras of observables. As most of the papers on Quantum Informations deal with matrix algebras one could get an impression that such approach is the proper one. However, it should be recognized that most of the extensions of this approach to more general situations has been plagued with the difficulties, e.g. even the standard canonical quantization demands infinite dimensional Hilbert spaces - although this feature of quantization is a part of the folklore of quantum physics the original credit might be give to Wintner [32] and Wieland [33]. To find the proper approach it is worth quoting Dirac's point of view on the structure of Quantum Theory:

"The states of dynamical variables have to be represented by mathematical quantities of different nature from those ordinarily used in physics. The new scheme becomes a precise physical theory when all the axioms and rules of the manipulations governing the mathematical quantities are specified and when in addition certain laws are laid down connecting physical facts with the mathematical formalism, so that from any given physical conditions equations between mathematical quantities may be inferred and vice versa." (see [11]) p. 15)

Following Dirac's recipe, the research done by Birkhoff, von Neumann, Jordan, Wigner, and Segal (see [5], [12], and [16]) led to the algebraic approach and algebras of operators came into the game. We emphasize that $M_n(\mathbb{C})$ and $\mathcal{B}(\mathcal{H})$ are basic examples of such algebras ($\mathcal{B}(\mathcal{H})$ stands for the set of all linear bounded operators on a Hilbert space $\mathcal{H}$). Consequently, an analysis of such fundamental concept as entanglement and its links to positive maps demands algebraic approach to Quantum Theory. In that way, the framework of our lecture which is based on operator algebras, should be clear and natural.

Let us turn to the positive maps. To begin with, let $\mathcal{A}$ and $\mathcal{B}$ denote $\mathcal{B}(\mathcal{H})$ and $\mathcal{B}(\mathcal{K})$ respectively, $\mathcal{A}_h = \{a \in \mathcal{A}; a = a^*\}$ - the set of all selfadjoint elements in $\mathcal{A}$, $\mathcal{A}^+ = \{a \in \mathcal{A}_h; a \geq 0\}$ - the set of all positive elements in $\mathcal{A}$, and $\mathcal{S}(\mathcal{A})$ the set of all states on $\mathcal{A}$, i.e. the set of all linear functionals φ on $\mathcal{A}$ such that $\varphi(1) = 1$ and $\varphi(a) \geq 0$ for any $a \in \mathcal{A}^+$. In particular

$$(\mathcal{A}_h, \mathcal{A}^+) \text{ is an ordered Banach space.}$$

We say that a linear map $\alpha : \mathcal{A} \to \mathcal{B}$ is positive if $\alpha(\mathcal{A}^+) \subset \mathcal{B}^+$. The set of all (linear, bounded) positive maps $\alpha : \mathcal{A} \to \mathcal{B}$ will be denoted by $\mathcal{L}^+(\mathcal{A}, \mathcal{B})$. Clearly, the set $\mathcal{L}^+(\mathcal{A}, \mathcal{B})$ is a convex set, i.e. the line segment

$$[\alpha, \alpha'] = \{\lambda\alpha + (1 - \lambda)\alpha'; \quad 0 \leq \lambda \leq 1\}$$

is entirely contained in $\mathcal{L}^+(\mathcal{A}, \mathcal{B})$ whenever its endpoints α and α' are in $\mathcal{L}^+(\mathcal{A}, \mathcal{B})$.

As it was mentioned in Introduction we will consider only bipartite systems i.e. we are interested in a subclass of composite systems. One of the basic tools needed for descriptions of such systems is the concept of tensor product. On the other hand, in his pioneering work on Banach spaces, Grothendieck[13] observed the links between tensor products and mapping spaces. As our characterization of PPT states stems from the mentioned links we summarize without proofs the relevant material on tensor products of Banach spaces.

Let X, Y be Banach spaces. We denote by $X \odot Y$ the algebraic tensor product of X and Y (algebraic tensor product of two vector spaces is defined as its *-algebraic structure when the factor spaces are *-algebras; so the topological questions are not considered) We consider the following norm on $X \odot Y$

$$\pi(u) = inf\{\sum_{i=1}^{n} ||x_i|| ||y_i|| : \quad u = \sum_{i=1}^{n} x_i \otimes y_i\}. \tag{1}$$

The norm π is called the projective norm. Furthermore, we denote by $X \otimes_\pi Y$ the completion of $X \odot Y$ with respect to the norm π and this Banach space will be referred as the projective tensor product of the Banach spaces X and Y.

Denote by $\mathcal{B}(X \times Y)$ the Banach space of bounded bilinear mappings from $X \times Y$ into the field of scalars with the norm given by $||B|| = sup\{|B(x, y)|; ||x|| \leq 1, ||y|| \leq 1\}$. Note (for all details see [27]), that with each bounded bilinear form $B \in \mathcal{B}(X \times Y)$ there is an associated operator $L_B \in \mathcal{L}((X, Y^*)$ defined by $< y, L_B(x) >= B(x, y)$. The mapping $B \to L_B$ is an isometric isomorphism between the spaces $\mathcal{B}(X \times Y)$ and $\mathcal{L}(X, Y^*)$. Hence, there is an identification

$$(X \otimes_\pi Y)^* = \mathcal{L}(X, Y^*), \tag{2}$$

such that the action of an operator $S : X \to Y^*$ as a linear functional on $X \otimes_\pi Y$ is given by

$$< \sum_{i=1}^{n} x_i \otimes y_i, S >= \sum_{i=1}^{n} < y_i, Sx_i > . \tag{3}$$

Let us make the following specification in the just given description of tensor products: $X \equiv \mathfrak{A}$ denotes a norm closed self-adjoint subspace of bounded operators on a Hilbert space $\mathcal{K}$ containing identity operator on $\mathcal{K}$. $Y \equiv \mathfrak{T}$ will denote the set of trace class operators on $\mathcal{B}(\mathcal{H})$. Moreover $\mathcal{B}(\mathcal{H}) \ni x \to x^t \in \mathcal{B}(\mathcal{H})$ denotes the transpose map of $\mathcal{B}(\mathcal{H})$ with respect to some orthonormal basis. The set of all linear bounded (positive) maps $\phi : \mathfrak{A} \to \mathcal{B}(\mathcal{H})$ will be denoted by $\mathcal{L}(\mathfrak{A}, \mathcal{B}(\mathcal{H}))$ ($\mathcal{L}(\mathfrak{A}, \mathcal{B}(\mathcal{H}))^+$ respectively). Finally, we denote by $\mathfrak{A} \odot \mathfrak{T}$ the algebraic tensor product of $\mathfrak{A}$ and $\mathfrak{T}$ and denote by $\mathfrak{A} \otimes_\pi \mathfrak{T}$ its projective tensor product, i.e. the Banach space closure under the projective norm defined by

$$||x|| = \inf\{\sum_{i=1}^{n} ||a_i|| ||b_i||_1 : x = \sum_{i=1}^{n} a_i \otimes b_i, \ a_i \in \mathfrak{A}, \ b_i \in \mathfrak{T}\}, \tag{4}$$

where $|| \cdot ||_1$ stands for the trace norm. Now, we can quote (see [28])

Lemma 2.1. *There is an isometric isomorphism* $\phi \to \tilde{\phi}$ *between* $\mathcal{L}(\mathfrak{A}, \mathcal{B}(\mathcal{H}))$ *and* $(\mathfrak{A} \otimes_\pi \mathfrak{T})^*$ *given by*

$$(\tilde{\phi})(\sum_{i=1}^{n} a_i \otimes b_i) = \sum_{i=1}^{n} Tr(\phi(a_i)b_i^t), \tag{5}$$

where $\sum_{i=1}^{n} a_i \otimes b_i \in \mathfrak{A} \odot \mathfrak{T}$.

Furthermore, $\phi \in \mathcal{L}(\mathfrak{A}, \mathcal{B}(\mathcal{H}))^+$ *if and only if* $\tilde{\phi}$ *is positive on* $\mathfrak{A}^+ \otimes_\pi \mathfrak{T}^+$.

To comment on this result we make

Remark 2.1.

(1) There is no restriction on the dimension of Hilbert space. In other words, this result can be applied to true quantum systems.

(2) In [29], Størmer showed that in the special case when $\mathfrak{A} = M_n(\mathbb{C})$ and $\mathcal{H}$ has dimension equal to n, the above Lemma is a reformulation of Choi result (cf. [6]-[8]).

Lemma 2.1 offers a characterization of the structure of positive maps (see [23]). However, this topic exceeds the scope of this lecture and we won't treat it. On the other hand, a modification of Lemma 2.1 (see the next Section) will be the starting point of our characterization of PPT (positive partial transposition) states.

3. PPT states

As before, let $\mathcal{B}(\mathcal{H})$ denote the set of all linear bounded operators on a Hilbert space $\mathcal{H}$. The set of all positive elements of $\mathcal{B}(\mathcal{H})$ is denoted by $\mathcal{B}(\mathcal{H})^+$ and the set of all states on $\mathcal{B}(\mathcal{H})$ is denoted by $\mathcal{S}_{\mathcal{B}(\mathcal{H})}$.

Let $\psi : \mathcal{B}(\mathcal{H}_1) \longrightarrow \mathcal{B}(\mathcal{H}_2)$ be a positive map. For $k \in \mathbb{N}$, we consider a map $\psi_k : M_k \otimes \mathcal{B}(\mathcal{H}_1) \longrightarrow M_k \otimes \mathcal{B}(\mathcal{H}_2)$ where M_k denotes the algebra of $k \times k$-matrices with complex entries and $\psi_k = I_{M_k} \otimes \psi$. We say that ψ is k-positive if the map ψ_k is positive. The map ψ is said *completely positive*(CP for short) when ψ is k-positive for every $k \in \mathbb{N}$.

Let us recall that for a Hilbert space $\mathcal{L}$, every normal (so *-weakly continuous) state $\phi(\cdot)$ on $\mathcal{B}(\mathcal{L})$ has the form of $\phi(A) = \mathrm{Tr}\,(\varrho A)$, where ϱ is a uniquely determined *density matrix*, i.e. an element of $\mathcal{B}(\mathcal{L})^+$ such that $\mathrm{Tr}\,\varrho = 1$.

To define PPT states and to discuss their structure we need some more notations. We fix orthonormal bases $\{e_i\}_i$ and $\{f_j\}_j$ of the spaces $\mathcal{H}$ and $\mathcal{K}$ respectively. For simplicity we will write $\mathcal{S}$, $\mathcal{S}_{\mathcal{H}}$, $\mathcal{S}_{\mathcal{K}}$ instead of $\mathcal{S}_{\mathcal{B}(\mathcal{H}) \otimes \mathcal{B}(\mathcal{K})}$, $\mathcal{S}_{\mathcal{B}(\mathcal{H})}$, $\mathcal{S}_{\mathcal{B}(\mathcal{K})}$, respectively. By $\mathcal{B}(\mathcal{L}) \ni a \mapsto a^t \in \mathcal{B}(\mathcal{L})$ we denote transposition maps on $\mathcal{B}(\mathcal{L})$ associated with basis $\{e_i\} \subseteq \mathcal{L}$. More precisely, such map is defined as follows. Firstly

$$Jx = J\sum_i (e_i, x)e_i = \sum_i \overline{(e_i, x)}e_i \tag{6}$$

gives definition of a conjugation on a Hilbert space $\mathcal{L}$, i.e. J is an antilinear map $J : \mathcal{L} \to \mathcal{L}$ such that for any $e, f \in \mathcal{L}$

$$(Jf, e) = (Je, f), \qquad J \cdot J = I. \tag{7}$$

Then, the transposition $a \to a^t$ is defined as

$$a \mapsto a^t = Ja^*J \equiv \tau_{\mathcal{H}}(a). \tag{8}$$

A positive map $\Psi : \mathcal{B}(\mathcal{H}) \longrightarrow \mathcal{B}(\mathcal{K})$ is called *decomposable* if there are completely positive maps $\Psi_1, \Psi_2 : \mathcal{B}(\mathcal{H}) \longrightarrow \mathcal{B}(\mathcal{K})$ such that $\Psi = \Psi_1 + \Psi_2 \circ \tau_{\mathcal{H}}$. Let $\mathcal{P}$, $\mathcal{P}_{\mathrm{C}}$ and $\mathcal{P}_{\mathrm{D}}$ denote the set of all positive, completely positive and decomposable maps from $\mathcal{B}(\mathcal{H})$ to $\mathcal{B}(\mathcal{K})$, respectively. Note that

$$\mathcal{P}_{\mathrm{C}} \subset \mathcal{P}_{\mathrm{D}} \subset \mathcal{P} \tag{9}$$

(see also [6]-[9]).

A state $\varphi \in \mathcal{S}$ is said to be *separable* if it can be approximated (in norm) by states of the form

$$\varphi_N = \sum_{n=1}^{N} \lambda_n \varphi_n^{\mathcal{H}} \otimes \varphi_n^{\mathcal{K}}$$

where $N \in \mathbb{N}$, $\varphi_n^{\mathcal{H}} \in \mathcal{S}_{\mathcal{H}}$, $\varphi_n^{\mathcal{K}} \in \mathcal{S}_{\mathcal{K}}$ for $n = 1, 2, \ldots, N$, λ_n are positive numbers such that $\sum_{n=1}^{N} \lambda_n = 1$, and the state $\varphi_n^{\mathcal{H}} \otimes \varphi_n^{\mathcal{K}}$ is defined as $\varphi_n^{\mathcal{H}} \otimes \varphi_n^{\mathcal{K}}(A \otimes B) = \varphi_n^{\mathcal{H}}(A)\varphi_n^{\mathcal{K}}(B)$ for $A \in \mathcal{B}(\mathcal{H})$, $B \in \mathcal{B}(\mathcal{K})$. The set of all separable states on the algebra $\mathcal{B}(\mathcal{H}) \otimes \mathcal{B}(\mathcal{K})$ is denoted by $\mathcal{S}_{\text{sep}}$. A state which is not in $\mathcal{S}_{\text{sep}}$ is called *entangled* or *non-separable*.

Finally, let us define the family of *PPT* (transposable) states on $\mathcal{B}(\mathcal{H}) \otimes \mathcal{B}(\mathcal{K})$

$$\mathcal{S}_\tau = \{\varphi \in \mathcal{S} : \varphi \circ (I_{\mathcal{B}(\mathcal{H})} \otimes \tau_{\mathcal{K}}) \in \mathcal{S}\}.$$

Note that due to the positivity of the transposition $\tau_{\mathcal{K}}$ every separable state φ is transposable, so

$$\mathcal{S}_{\text{sep}} \subset \mathcal{S}_\tau \subset \mathcal{S}. \tag{10}$$

As PPT states are "dual" to decomposable maps (see [24]) we need an adaptation of Lemma 2.1 for CP and co-CP maps. In [25], it was showed:

Lemma 3.1. *(1) Let* $\mathcal{B}[\,\mathcal{B}(\mathcal{H}), \mathcal{B}(\mathcal{K})_*]$ *stand for the set of all linear, bounded, normal (so weakly* $*$*-continuous) maps from* $\mathcal{B}(\mathcal{H})$ *into* $\mathcal{B}(\mathcal{H})_*$*. There is an isomorphism* $\psi \longmapsto \Psi$ *between* $\mathcal{B}[\,\mathcal{B}(\mathcal{H}), \mathcal{B}(\mathcal{K})_*]$ *and* $(\mathcal{B}(\mathcal{H}) \otimes_\pi \mathcal{B}(\mathcal{K}))_*$ *given by*

$$\Psi\left(\sum_i a_i \otimes b_i\right) = \sum_i Tr_{\mathcal{K}}\psi(a_i)\, b_i^t, \quad a_i \in \mathcal{B}(\mathcal{H}), \ b_i \in \mathcal{B}(\mathcal{K}). \tag{11}$$

The isomorphism is isometric if Ψ *is considered on* $\mathcal{B}(\mathcal{H}) \otimes_\pi \mathcal{B}(\mathcal{K})$*. Furthermore* Ψ *is positive on* $(\mathcal{B}(\mathcal{H}) \otimes \mathcal{B}(\mathcal{K}))^+$ *iff* ψ *is complete positive.*

(2) There is an isomorphism $\phi \longmapsto \Phi$ *between* $\mathcal{B}[\,\mathcal{B}(\mathcal{H}), \mathcal{B}(\mathcal{K})_*]$ *and* $(\mathcal{B}(\mathcal{H}) \otimes_\pi \mathcal{B}(\mathcal{K}))_*$ *given by*

$$\Phi\left(\sum_i a_i \otimes b_i\right) = \sum_i Tr_{\mathcal{K}}\phi(a_i)\, b_i, \quad a_i \in \mathcal{B}(\mathcal{H}), \ b_i \in \mathcal{B}(\mathcal{K}). \tag{12}$$

The isomorphism is isometric if Φ *is considered on* $\mathcal{B}(\mathcal{H}) \otimes_\pi \mathcal{B}(\mathcal{K})$*. Furthermore* Φ *is positive on* $(\mathcal{B}(\mathcal{H}) \otimes \mathcal{B}(\mathcal{K}))^+$ *iff* ϕ *is complete co-positive.*

Lemma 3.1 adduce a reason to define the so called *entanglement mappings* [3], see also [4], [25], and [26]. To this end we will follow Belavkin-Ohya scheme (B-O, for short). Let us fix a normal state ω on $\mathcal{B}(\mathcal{H}) \otimes \mathcal{B}(\mathcal{K})$, i.e. a density matrix ϱ describing a composite state ω is fixed. Denote by $\sum_i \lambda_i |e_i\rangle\langle e_i|$ its spectral decomposition. Finally we define the embedding $T_\zeta : \mathcal{K} \to \mathcal{H} \otimes \mathcal{K}$ by

$$T_\zeta \eta = \zeta \otimes \eta \tag{13}$$

where $\zeta \in \mathcal{H}$, $\eta \in \mathcal{K}$. The entanglement operator H

$$H : \mathcal{H} \to \mathcal{H} \otimes \mathcal{K} \otimes \mathcal{K} \tag{14}$$

is given by the formula:

$$H\zeta = \sum_i \lambda_i^{\frac{1}{2}} \left(J_{\mathcal{H}\otimes\mathcal{K}} \otimes T_{J_{\mathcal{H}}\zeta}^* \right) e_i \otimes e_i \tag{15}$$

where $J_{\mathcal{H}\otimes\mathcal{K}}$ is a complex conjugation defined by $J_{\mathcal{H}\otimes\mathcal{K}} f \equiv J_{\mathcal{H}\otimes\mathcal{K}}(\sum_i (e_i, f) e_i) = \sum_i \overline{(e_i, f)} e_i$ where $\{e_i\}$ is any CONS (complete orthonormal system) extending (if necessary) the orthogonal system $\{e_i\}$ determined by the spectral resolution of ρ. ($J_{\mathcal{H}}$ is defined analogously with the spectral resolution given by $H^* H$). The entanglement mapping ϕ is defined as

$$\phi(b) = (H^* (1 \otimes b) H)^t = J_{\mathcal{H}} H^* (1 \otimes b)^* H J_{\mathcal{H}}. \tag{16}$$

The properties of the entanglement mapping and its dual were studied in [25]. In particular, one has (see [25])

Proposition 3.1. *The entanglement mapping*
(i) $\phi^* : \mathcal{B}(\mathcal{H}) \to \mathcal{B}(\mathcal{K})_*$ has the following explicit form

$$\phi^*(a) = Tr_{\mathcal{H}\otimes\mathcal{K}} H a^t H^* \tag{17}$$

(ii) The state ω on $\mathcal{B}(\mathcal{H} \otimes \mathcal{K})$ can be written as

$$\omega(a \otimes b) = Tr_{\mathcal{H}} a\phi(b) = Tr_{\mathcal{K}} b\phi^*(a) \tag{18}$$

where ϕ was defined in (16).

Proposition 3.1 and the fact that, by definition, any PPT state composed with partial transposition is again a state lead to

Corollary 3.1. *PPT states are completely characterized by entanglement mappings ϕ^* which are both CP and co-CP.*

Consequently, for an arbitrary but fixed normal state, there exists the explicit construction of an entanglement map. If this mapping, for given state, is both CP and co-CP (what is verifiable) the considered state is PPT. Thus we got *operational and complete characterization of the set of PPT states*.

4. Measures of entanglement

As it was mentioned, it is widely known that quantum entanglement is one of the basic causes of fundamental differences between quantum and classical mechanics (see [15]). Formally, and in a rather non-constructive way, an entangled state ϱ was defined through the non-existence of a convex decomposition (see [31] and the previous Section) i.e. there does not exist any convex combination $\varrho_0 = \sum_i p_i \varrho_i^1 \otimes \varrho_i^2$ such that ϱ would be equal to ϱ_0.

Here we wish to examine the entanglement in more constructive way. To this end we review our recent results on quantifying entanglement. It should be stressed that there are other approaches e.g. see [1], [15] and the references given there. However, our approach stems from:

(1) the algebraic approach outlined in the first part of Section 2,
(2) the observation that correlations contained in entangled states can not be expressed in classical terms (see [19], [20]),
(3) two hallmarks of quantumness of a system: the first one is saying that, contrary to classical mechanics, the set of all states of a quantum system is not a simplex. The second one can be formulated as follows: a restriction of a pure state of a composite system to one of its subsystems does not need to be pure.

To describe our results, again, we need some preparation. We start with a brief exposition of Tomita-Takesaki theory (see [17], and [25]). It is well known that this theory is in the heart of algebraic formalism (for a comprehensive account of this theory addressed to physicists we refer Haag's book,[16] while the mathematical description can be found in [2], [5], and [34]).

Let $\mathcal{H}$ be a Hilbert space. Define $\omega \in \mathcal{S}_{\mathcal{B}(\mathcal{H})}$ as $\omega(a) = \operatorname{Tr} \varrho a$, where ϱ is an invertible density matrix, i.e. the state ω is a faithful one. As it is accepted in the physical literature we will frequently identify the state ω with its density matrix ϱ. Denote by $(\mathcal{H}_\pi, \pi, \Omega)$ the GNS triple associated with $(\mathcal{B}(\mathcal{H}), \omega)$. Then, one has:

- $\mathcal{H}_\pi$ is identified with $\mathcal{B}(\mathcal{H})$ where the inner product $(\cdot,\cdot)$ is defined as $(a,b) = \operatorname{Tr} a^* b$, $a, b \in \mathcal{B}(\mathcal{H})$;
- With the above identification: $\Omega = \varrho^{1/2}$;
- $\pi(a)\Omega = a\Omega$; $\omega(a) = (\Omega, \pi(a)\Omega)$.
- The modular conjugation J_m is the hermitian involution: $J_\mathrm{m} a \varrho^{1/2} = \varrho^{1/2} a^*$;
- The modular operator Δ is equal to the map $\varrho \cdot \varrho^{-1}$, where the essential domain of Δ is given by $\{\pi(a)\Omega : a \in \mathcal{B}(\mathcal{H})\}$.
- The natural cone $\mathcal{P}$ is given by

$$\mathcal{P} = \operatorname{closure} \left\{ \Delta^{\frac{1}{2}} a \varrho^{1/2} :\ a \geq 0,\ \ a \in \mathcal{B}(\mathcal{H}) \right\}. \tag{19}$$

The significance of the natural cone $\mathcal{P}$ for description of states of a quantum system follows from the fact that there is the one-to-one correspondence between the natural cone $\mathcal{P}$ and the set of normal states on $\pi(\mathcal{B}(\mathcal{H}))$ described in [10], [5, Theorem 2.5.31], i.e. such that

$$\omega_\xi(a) = (\xi, a\xi), \quad a \in \mathcal{B}(\mathcal{H}), \quad \xi \in \mathcal{P}. \tag{20}$$

Let us apply the just given scheme to the description of $\mathcal{S}_\mathrm{sep}$ and $\mathcal{S}_\tau$. Firstly, we will present a characterization of $\mathcal{S}_\mathrm{sep}$. We consider a composite system $A + B$ where a subsystem $i = A, B$ is described by $(\mathcal{B}(\mathcal{K}_i), \mathcal{P}_i, \varrho_i)$ where $\mathcal{P}_i$ denotes the natural cone associated with $(\mathcal{B}(\mathcal{K}_i), \varrho_i)$ (cf. [18] and the just presented description of Tomita-Takeski theory). In [18], using Tomita-Takesaki approach, we have obtained

Proposition 4.1. *There is a one-to-one correspondence between the set of normalized vectors in $\mathcal{P}_A \otimes \mathcal{P}_B$ and the set of all separable states, where*

$$\mathcal{P}_A \otimes \mathcal{P}_B \equiv \operatorname{closure}\{\sum_k a_k x_k^{(1)} \otimes x_k^{(2)}, a_k \geq 0, \sum_k a_k = 1, x_k^{(i)} \in \mathcal{P}_i\}. \tag{21}$$

Secondly, we wish to extend this result and to get an analogous characterization of PPT states. Assume for simplicity that the Hilbert space associated with the second subsystem is finite dimensional i.e. $\mathcal{H} \equiv \mathbb{C}^n$. Then the natural cone $\mathcal{P}_n$ associated with the composite system $(\mathcal{A} \otimes \mathcal{B}, \varrho_1 \otimes \varrho_2)$ with faithful states ϱ_1, ϱ_2 may be realized as

$$\mathcal{P}_n = \overline{\{\Delta_n^{1/4}[a_{ij}]\Omega :\ \ [a_{ij}] \in M_n^\pi(\mathcal{A})^+\}} \tag{22}$$

where $\pi(\mathcal{A} \otimes \mathcal{B}(\mathbb{C}^n)) \equiv M_n^\pi(\mathcal{A})$, $\Omega = \Omega_1 \otimes \Omega_2$, and obviously, $\overline{X}$ stands for the closure of X.

We will also need (see [17]) the notion of the "transposed cone" $\mathcal{P}_n^\tau \equiv (\mathbf{I} \otimes U)\mathcal{P}_n$, where τ is transposition on $M_n(\mathbb{C})$ while the operator U is defined as follows

$$U = \sum_{ij} |E_{ji}\rangle\langle E_{ij}| \tag{23}$$

We are using the following identification: for the basis $\{e_i\}_i$ in $\mathbb{C}^n$ consisting of eigenvectors of ϱ_2 $(\omega_2(\cdot) = Tr\{\varrho_2\cdot\}$, we have the basis $\{E_{ij} \equiv |e_i\rangle\langle e_j|\}_{ij}$ in the GNS Hilbert space associated with $(\mathcal{B}(\mathbb{C}^n), \omega_2)$ with U defined in terms of that basis). Note that in the same basis one has the identification of $\mathcal{B}(\mathbb{C}^n)$ with $M_n(\mathbb{C})$. We are now in a position to define the transposed cone $\mathcal{P}_n^\tau$:

$$\mathcal{P}_n^\tau = \overline{\Delta^{1/4}\{[a_{ji}]\Omega : [a_{ij}] \in M_n^\pi(\mathcal{A})^+\}}. \tag{24}$$

where $\Omega \equiv \Omega_1 \otimes \Omega_2$. In [25] we have proved

Theorem 4.1. *Assume that both subsystems are finite dimensional. Then, there is one-to-one correspondence between the set of PPT states and $\mathcal{P}_n \cap \mathcal{P}_n^\tau$.*

To comment Theorem 4.1 we note:

(1) Theorem 4.1 follows from the characterization of $\mathcal{P}_n \cap \mathcal{P}_n^\tau$ (for details and for the discussion about a possible extension of this result to true quantum systems we refer to [17]).
(2) As U is nontrivial the inclusion $\mathcal{P}_n \cap \mathcal{P}_n^\tau \subseteq \mathcal{P}_n$ should be, in general, the proper one.
(3) As $\mathcal{P}_n^\tau$ and $\mathcal{P}_n$ contain $\mathcal{P}_A \otimes \mathcal{P}_B$, PPT states which are not separable are characterized by vectors in $\mathcal{P}_n \cap \mathcal{P}_n^\tau \setminus \mathcal{P}_A \otimes \mathcal{P}_B$. Thus, Theorem 4.1 gives a quite effective recipe for a construction of PPT state which is not a separable one (see [21]).
(4) Similarly, non-PPT states are characterized by vectors in $\mathcal{P}_n \setminus \mathcal{P}_n \cap \mathcal{P}_n^\tau$. Again, this gives a recipe for a construction of non-PPT states.
(5) The above characterization of PPT states can be considered as dual approach to the characterization obtained by Belavkin and Ohya (see [3], [4], [25], and [26]).

In [19], [20], working within the algebraic approach, we have introduced the degree of quantum correlations. To take into account the discussed hallmarks of quantumness of a system (see the beginning of this Section) we have used the so called decomposition theory in operator algebras. In other words we have worked in the Heisenberg picture. Surprisingly, a repetition of these ideas, but now in the context of Hilbert spaces, offers a great simplification. In other words, working in the Schroedinger picture, one can employ the geometry of Hilbert spaces and exactly this geometry leads to the simplification. We begin with (cf. [21], [25])

Definition 4.1. Let ξ be a vector in the natural cone $\mathcal{P}_n$ corresponding to a normal state of a composite system $1 + 2$ (cf. Proposition 4.1 and Theorem 4.1). Then

(1) Degree of entanglement (or *quantum correlations*) is given by:

$$D_e(\xi) = inf\{||\xi - \eta||; \eta \in \mathcal{P}_1 \otimes \mathcal{P}_2\} \qquad (25)$$

(2) Degree of genuine entanglement (or *genuine quantum correlations*) is defined as

$$D_{ge}(\xi) = inf\{||\xi - \eta||; \eta \in \mathcal{P} \cap \mathcal{P}^\tau\} \qquad (26)$$

We will briefly discuss the geometric idea behind this definitions. The key to the argument is convexity (in Hilbert spaces). Namely, we observe

(1) $\mathcal{P}_1 \otimes \mathcal{P}_2 \subset \mathcal{P}$ is a convex subset,
(2) $\mathcal{P} \cap \mathcal{P}^\tau \subset \mathcal{P}$ is a convex subset,
(3) the theory of Hilbert spaces says: there exists a unique vector (representing a separable state) $\xi_0 \in \mathcal{P}_1 \otimes \mathcal{P}_2$, such that $D_e(\xi) = ||\xi - \xi_0||$,
(4) analogously, there exists a unique vector (representing a PPT state) $\eta_0 \in \mathcal{P} \cap \mathcal{P}^\tau$, such that $D_{ge}(\xi) = ||\xi - \eta_0||$.

The important point to note here is that we used the well known property of convex subsets in a Hilbert space $\mathcal{H}$: fix a vector $\xi \in \mathcal{H}$, a closed convex subset W in a Hilbert space $\mathcal{H}$ contains the unique vector with the smallest distance from ξ. This ensures the existence of vectors ξ_0 and η_0 introduced in (3) and (4) respectively. Furthermore, if we are considering a vector $\xi \in \mathcal{P} \setminus \mathcal{P}_1 \otimes \mathcal{P}_2$ representing an entangled state, then $D_e(\xi)$ measures how far this state is from "the classical world". Thus, $D_e(\xi)$ describes how big amount of quantum correlations is contained in the state. Analogously, for a vector $\xi \in \mathcal{P} \setminus \mathcal{P} \cap \mathcal{P}^\tau$, $D_{ge}(\xi)$ measures how far this state is from "a world" having classical and weak quantum correlations only (weak as a PPT state can contain some quantum correlations). This explains the name of $D_{ge}(\xi)$.

Having such classification of entanglement one can ask for its effectiveness. We have shown in [25] that using Hilbert space geometry and Tomita's theory we were able to estimate $D_e(\xi)$ for concrete states (equivalently, for fixed vectors ξ in the natural cone). Moreover, degrees $D_e(\xi)$ and $D_{ge}(\xi)$ exhibit all expected properties of "well-designed" measures of entanglement.

Concluding, for an entangled (non-PPT state) we are able to find *the best approximation* among separable states (PPT states, respectively). Moreover, this approach offers a classification of entanglement (genuine entanglement, respectively). More detailed discussion appeared in [25].

Acknowledgments

The author would like to acknowledge the partial supports of the grant number N N202 208238 of Polish Ministry of Science and Higher Education is gratefully acknowledged.

Bibliography

1. G. Adler, et al., *Quantum Information. An Introduction to basic theoretical concepts and experiments.*, Springer Tracts in Modern Physics **173**, 2001.
2. H. Araki, Some properties of modular conjugation operator of a von Neumann algebra and non-commutative Radon-Nikodym theorem with a chain rule, *Pac, J. Math.* **50** (1974), 309.
3. V. P. Belavkin, M. Ohya, Quantum entropy and information in discrete entangled states, *Infinite analysis, quantum probability and related topics*, 4 137 (2001).
4. V. P. Belavkin, M. Ohya, Quantum entanglement and entangled mutual entropy, *Proc. Royal. Soc. London A*, **458** 209 (2002).
5. Bratteli, O. and Robinson, D. W., *Operator Algebras ans Quantum Statistical Mechanics I*, Springer Verlag, 1987.
6. Choi, M.-D., Positive linear maps, *Proc. Sympos. Pure. Math.* **38**, 583-590 (1982).
7. M.-D. Choi, Positive semidefinite biquadratic forms, *Lin. Alg. Appl.* **12** 95–100 (1975).
8. Choi, M.-D., Completely Positive Maps on Complex Matrices *Lin. Alg. Appl.* **10** (1975), 285–290.
9. Choi, M.-D., Some assorted inequalities for positive linear maps on C^*-algebras, *J. Operator Th.* **4** (1980), 271–285.
10. A. Connes, Characterization des espaces vectoriels ordonnés sous-jacents aux algébres de von Neumann, *Ann. Inst. Fourier, Grenoble* **24** 121–155 (1974).
11. P. A. M. Dirac, *The Principles of Quantum Mechanics*, Oxford, third edition, 1948.
12. G. Emch, *Algebraic Methods in Statistical Mechanics and Quantum Field Theory*, Wiley-Interscience, 1972.

13. A. Grothendieck, Products tensoriels topologiques et espaces nuclearies, *Memoirs of the American Mathematical Society*, **16**, Providence.

14. Horodecki, M., Horodecki, P. and Horodecki R., Separability of mixed states: necessary and sufficient conditions, *Phys. Lett. A* **223**, 1–8 (1996).

15. R. Horodecki, P. Horodecki, M. Horodecki, K. Horodecki, Quantum entanglement, *Rev. Mod. Phys* **81**, 865–942 (2009).

16. R. Haag, *Local Quantum Physics; Fields, Particles, Algebras*, second edition, Springer Verlag 1996.

17. L. E. Labuschagne, W. A. Majewski, M. Marciniak, On k-decomposability of positive maps, *Expo. Math.* **24**, 103–125 (2006); math-ph/0306017.

18. W. A. Majewski, Separable and entangled states of composite quantum systems - rigorous description, *Open Sys. & Inf. Dyn.* **6**, 79–86 (1999).

19. W. A. Majewski, "On entanglement of states and quantum correlations",in *Operator algebras and Mathematical Physics*, Eds. J. M. Combes, J. Cuntz, G. A. Elliott, G. Nenciu, H. Siedentop, S. Stratila; pp. 287–297, *Theta*, Bucharest, 2003. e-print, LANL math-ph/0202030.

20. W. A. Majewski, "On quantum correlations and positive maps", *Lett. Math. Physics,* **67**, 125–132 (2004); e-print, LANL math-ph/0403024.

21. W. A. Majewski, Measures of entanglemen - a Hilbert space approach, *Quantum Prob. and White Noise Analysis* **XXIV** pp. 127–138, 2009.

22. W. A. Majewski, Positive maps, entanglement and all that; some old and new problems, arXiv:quant-ph/0411043.

23. W. A. Majewski, On the structure of positive maps; finite dimensional case, arXive:math-ph/1005.3949.

24. W. A. Majewski, M. Marciniak, On a characterization of positive maps, *J. Phys. A* **34** 5863–5874 (2001).

25. W. A. Majewski, T. Matsuoka, M. Ohya, Characterization of partial positive transposition states and measures of entanglement, *J. Math. Phys.* 50 113509 (2009).

26. T. Matsuoka, Some characterization of PPT states and their relations, *Quantum Prob. and White Noise Analysis* **XXIV** pp. , 2009.

27. R. A. Ryan, *Introduction to tensor products of Banach spaces*, Springer Verlag, 2002.

28. E. Størmer, Extension of positive maps, *J. Funct. Analysis,* **66** (1986), 235–254.

29. E. Størmer, Cones of positive maps, *Contemporary Mathematics* **62** (1987), 345–356.

30. E. Størmer. Positive linear maps on operator algebras, *Acta Math.* **110** (1963) 233–278.

31. R. F. Werner, Quantum states with Einstein-Podolsky-Rosen correlations addimiting a hidden-variable model, *Phys. Rev. A*, **40**, 4277 (1989).

32. A. Wintner, The unboundedness of quantum mechanical matrices, *Phys. Rev.*, **71**, 737–739 (1947).

33. H. Wieland, Über die Unbeschränktheit der Schrödingerschen Operatoren der Quantenmmechanik, *Math. Ann.*, **121**, 21 (1949).

34. Takesaki M., *Tomita's theory of modular Hilbert algebras and its applications*, Lecture Notes in Mathematics **128** Springer Verlag, Berlin, 1970.

Entanglement via nilpotent polynomials

Aikaterini Mandilara

Vladimir M. Akulin

Quantum Information and Communication, École Polytechnique, CP 165/59,
Université Libre de Bruxelles, 1050 Brussels, Belgium

Laboratoire Aime Cotton, CNRS, Campus d'Orsay, 91405, Orsay, France

This paper is an attempt to give a pedagogical presentation of the original work by Lorenza Viola, Andrei Smilga, and ourselvs published in Physical Review **A** 74, 022331 (2006). The readers are invited to consult that paper for more details as well as for the history of the entanglement concept and the relevant bibliography.

Keywords: entanglement; tanglemeters; nilpotent polynomials.

1. Some group-theory aspects of entanglement

In contrast to the majority of the characteristics of physical systems, quantum entanglement is not something that exists as an actor in reality and governs dynamics of natural processes. It is rather a concept convenient for description and understanding of the relation among an entire system and its parts, than a physical quantity. The description of entanglement relies on a specific partition of the composite physical system sketched in Fig.1, which illustrates that the system can often be decomposed into a number of subsystems in many different ways, each of the subsystems possibly being a composite system by itself. In order to avoid an ambiguity related to such a situation, we have to introduce a few definitions. Given a partition of the composite system into n subsystems, we call each of them an "element" and characterize it by a single, possibly collective, quantum number. Thus, the composite system is a collection of the elements. We call this collection an "assembly" in order to avoid confusion with an "ensemble", which is usually understood as a set of all possible realizations of a many-body system with an associated probability distribution over these realizations. The i-th element is assumed to have Hilbert space of dimension d_i. Qubit, qutrit,

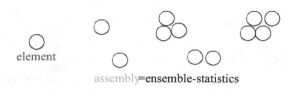

element

assembly=ensemble-statistics

Figure 1. First step for the entanglement characterization is to choose a partition: An assembly consists of elements each of which can be a composite system by itself.

and qudit are widely used names for two-, three-, and d-level elements (with $d_i = 2$, $d_i = 3$, and $d_i = d$, respectively).

The term "entanglement" has a transparent qualitative meaning: A pure state of an assembly is entangled with respect to a chosen partition when its state vector cannot be represented as a direct product of state vectors of the elements. This notion can also be extended to generic mixed quantum states, whereby entanglement is defined by the inability to express the assembly density operator as a probabilistic combination of direct products of the density operators of the elements, but this issue will not be addressed here in detail.

We will focus at the question: how to quantitatively characterize multipartite entanglement for an assembly of many elements in a pure state? This question can be easily answered for the bipartite setting $n = 2$ that is, for an assembly consisting just of two distinguishable elements A and B, each of arbitrary dimension. In this case, the von Neumann $\mathrm{Tr}[\rho_A \log \rho_A]$ or linear $\mathrm{Tr}[\rho_A^2]$ entropies based on the reduced density operator of either element, e.g., $\rho_A = \mathrm{Tr}_B[|\Psi\rangle \langle \Psi|]$, may be chosen as entanglement measures for a pure state $|\Psi\rangle$ of the assembly. However, already for the tripartite case $n = 3$, characterizing and quantifying entanglement becomes much harder. In fact, there are not one, but many different characteristics of entanglement and, apart from the question "How much?", one has also to answer the question "In which way" are different elements entangled? Therefore the construction of appropriate measures is not unique, and is mostly dictated by convenience.

The main requirement imposed on the construction is that the entanglement characteristics should remain invariant with respect to the local operations, that are quantum transformations one applies to each element of the assembly individually. Note that the local operations can be either unitary or simply invertible, depending on the physical situation under consideration. The group theory offers a natural language for the situation: local operations form a subgroup of all possible transformations of the as-

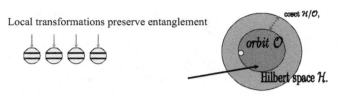

Figure 2.   Local transformations move the state vector along an orbit. Coset dimention of the subgroup of the local transformations (dimensionality of the space orthoganal to the orbit) determines how many characteristics of the entanglement are required.

sembly state vector. Under the action of the subgroup, the state vector of the system undergoes changes, still remaining within an orbit, that is a subset $\mathcal{O}$ of the overall Hilbert space $\mathcal{H}$. The dimension of the coset $\mathcal{H}/\mathcal{O}$, yields the number of independent invariants identifying a generic orbit (see Fig.2). Still, there are singular classes of orbits that require special consideration. One may want to directly construct a set of these invariants under local operations. When dealing with an assembly of k-qubits, this task can be accomplished in a regular and straightforward way. One considers the state vector of the assembly $\psi_{n_1 n_2 \ldots n_k}$ in the representation where each index $n_i = 0, 1$ denotes the state of the qubit i, and its dual vector $\psi^{n_1' n_2' \ldots n_k'}$ given by the convolution

$$\psi^{n_1' n_2' \ldots n_k'} = \epsilon^{n_1 n_1'} \epsilon^{n_2 n_2'} \ldots \epsilon^{n_k n_k'} \psi_{n_1 n_2 \ldots n_k}, \tag{1}$$

with the antisymmetric tensors $\epsilon^{ii'}$ of ranks 2. The required invariants are various convolutions of $\psi_{n_1 n_2 \ldots n_k}$ and $\psi^{n_1' n_2' \ldots n_k'}$. The explicit form of these invariants depends critically on the number of elements in the assembly. It has a simple form for 2-qubit assembly

$$I = \psi_{nm} \psi^{nm} = \psi_{00} \psi_{11} - \psi_{01} \psi_{10}. \tag{2}$$

For a 3-qubit assembly the situation gets more complicated – five independent local invariants exist, namely three real numbers

$$
\begin{aligned}
I_1 &= \psi_{kij} \psi^{*pij} \psi_{pmn} \psi^{*kmn} \ , \\
I_2 &= \psi_{ikj} \psi^{*ipj} \psi_{mpn} \psi^{*mkn} \ , \\
I_3 &= \psi_{ijk} \psi^{*ijp} \psi_{mnp} \psi^{*mnk} \ ,
\end{aligned}
\tag{3}
$$

and the real and the imaginary part of a complex number,

$$I_4 + \mathrm{i} I_5 = \psi_{ijk} \psi^{ijp} \psi_{mnp} \psi^{mnk}. \tag{4}$$

The quantity $2|I_4 + iI_5|$ is also known by the name residual entanglement or 3-tangle τ.

Similar invariants can still be found for a 4-qubit system. However, with increasing n, the explicit form of the invariants becomes less and less tractable and convenient for practical use, since for $n > 4$ even verification of functional independence of the chosen invariants may become a computational challenge. Moreover, no explicit physical meaning can be attributed to such invariants, while for the elements larger than qubits no analog of Eq.(1) for the invariants exists to our knowledge. Briefly, explicit construction of the orbit invariants becomes unpractical for assemblies of more than a few qubits.

2. Canonic states of an assembly

Fortunately apart of the invariants, there exists another way to identify an orbit, which the entanglement characterization can rely on. Every orbit can be represented by a "marker", which is one of the orbit state vectors selected according a certain rule. In order to unambiguously attribute a marker to each orbit, we specify a canonic form of an entangled assembly state. To this end, we first identify a reference state $|O\rangle$ as a direct product of the element states (see Fig.3). The latter can be chosen in an arbitrary way, but for qubits the choice $|0\rangle$, with the lowest energy level occupied, is the most natural. Thus, the reference state reads $|O\rangle = |0, \ldots 0\rangle$. Drawing parallels with quantum field theories and spin systems, we will call $|O\rangle$ the "ground" or "vacuum" state.

By applying local unitary operations to a generic quantum state $|\Psi\rangle$, we can bring it into the "canonic form" $|\Psi_c\rangle$ corresponding to the maximum possible population of the reference state $|O\rangle$. In other words, we apply a direct product $U_1 \otimes \ldots \otimes U_n$ of local transformations to the state vector $|\Psi\rangle$, and choose the transformation parameters to maximize $|\langle O| U_1 \otimes \ldots \otimes U_n |\Psi\rangle|^2$. For an assembly of qubits, the local transformation $U(n) = U_1 \otimes \ldots \otimes U_n$ satisfying this requirement

$$|\langle O| U_1 \otimes \ldots \otimes U_n |\Psi\rangle|^2 = \max$$

can be seen to be unique up to phase factors multiplying the upper states of each qubit. These factors can further be specified by the requirement that all amplitudes of the states where all but one qubits are in the upper state are real.

However, for a larger size of the elements in the assembly the requirement imposed to the phases is not sufficient for an unambiguous definition of the canonic state. Nevertheless, one can consider subgroup $U(n, -1) = U'_1 \otimes \ldots \otimes U'_n$ of local transformations acting exclusively in the

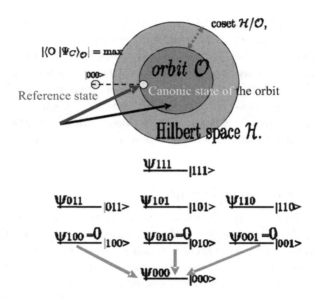

Figure 3.   Canonic state determined as the orbit state closest to a chosen product state (say vacuum state) serves as the orbit marker. Within the orbit, it corresponds to the maximum possible population of the vacuum state, as it is shown at the level scheme for a 3-qubit assembly. Canonic form of the entangled state for three qubits. By a local transformation the amplitudes of the lowest excited states *and* the phases of the second highest exited states are set to zero. The amplitude of the lowest state is taken as the common factor determining the normalization and the global phase.

subspace of on the element states other than the ground state. Application of these transformations will not change the projection of the assembly state vector at the reference state. In order to uniquely define the canonic state we have to apply a similar procedure to the excited states of the elements, namely choose another product reference state (say $|1\rangle = |1,\ldots 1\rangle$) that is orthogonal to all the states where at least one of the elements is in the ground state and maximize the projection of the assembly state vector to the state $|1\rangle$ within $U(n, -1)$.

Note that if some of the elements are qubits, the procedure of maximizing the overlap with the state $|1\rangle$ involves exclusively the elements other than these qubits. After having been performed, it allows one to uniquely identify the canonic state for qubits and qutrits entering the assembly. Application of a similar procedures to the elements of yet higher dimensions yields at the end a uniquely defined canonic state of the entire assembly. In Fig.3 we illustrate this for the assemblies of qutrits (a) for the local $SU(3)$

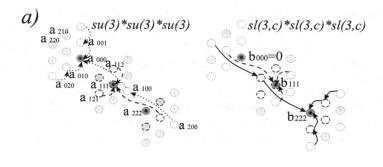

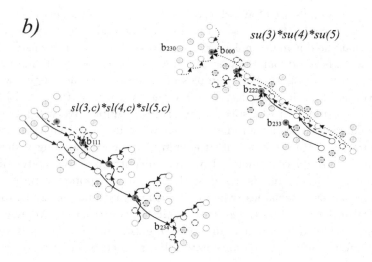

Figure 4. Canonic states for the elements other than quibits require more than one reference product state. One can see this at the example of an assembly of three qutrits (a) and an assembly comprising three different qudits of $d = 3, d = 4$, and $d = 5$ levels

and $SL(3)$ transformations, and for an assembly comprising three elements of different size (b). For an assembly shown in Fig.4(b) of a five-level system, a four-level system, and a qutrit, one should consecutively maximize:

(i) the population of the state $|0, 0, 0\rangle$ by the transformations from $SU(5) \otimes SU(4) \otimes SU(3)$,

(ii) the population of the state $|1, 1, 1\rangle$ by the transformations from $SU(4) \otimes SU(3) \otimes SU(2)$,

(iii) the population of the state $|2, 2, 2\rangle$ by the transformations from $SU(3) \otimes SU(2)$,

(iv) the population of the state $|3, 3, 2\rangle$ by the transformations from the remaining $SU(2)$ mixing the 3-d and the 4-th excited states of the five-level system.

Choice of the canonic states uniquely specifies the parameters of local transformations and we thus left only with the number of parameters of the state vector required exclusively for the entanglement characterization.

3. Extensive characteristics based on nilpotent polynomials: nilpotential and tanglemeter

Given the canonic state of an orbit, one still needs to characterize entanglement corresponding to this state in an economic and exhaustive way. We believe that the optimum choice for the purpose is so called extensive characteristics. Thermodynamic potentials linearly scaling with the number of particles in the system offer examples of extensive characteristics widely employed in statistical physics. The free energy given by the logarithm of the partition function is a specific important example. We will introduce similar characteristics for entangled states in such a way that their values for a product state coincide with the sum of the corresponding values for unentangled groups of elements. However, in contrast to thermodynamical potentials, this technique is based on the notion of nilpotent variables and functions of these variables, whence the logarithm function transforming products into sums plays the central role in the construction. An algebraic variable x is called nilpotent if an integer n exists such that $x^n = 0$. In the case of qubits, these variables are naturally provided by creation operators

$$\sigma_i^+ = \begin{pmatrix} 0 & 1 \\ 0 & 0 \end{pmatrix} .$$

Evidently, $(\sigma_i^+)^2 = 0$, since the same quantum state cannot be created twice.

Our approach is based on three main ideas: *(i)* We express the state vector of the assembly in terms of a polynomial of creation operators for elements applied to a reference product state. *(ii)* Rather than working with the polynomial of nilpotent variables describing the state, we consider its logarithm, which is also a nilpotent polynomial. Due to the important role that this quantity will play throughout the development, we call this quantity the "nilpotential" henceforth, by analogy to thermodynamic potentials. *(iii)* The nilpotential is not invariant under local transformations, being different in general for different states in the same orbit. We therefore consider nilpotential of the canonic state of an orbit, which contains complete information about the entanglement in the assembly. We therefore call this quantity the "tanglemeter". The latter is, by construction, extremely convenient as an extensive orbit marker: the tanglemeter for an assembly consisting of several unentangled groups of elements equals the sum of tanglemeters of these groups.

Let us explain these ideas, in the simplest example of n qubits subject to the $su(2)_1 \oplus \ldots \oplus su(2)_i \ldots \oplus su(2)_n$ Lie algebra of local transformations. As a reference state, we choose the Fock vacuum that is, the state $|O\rangle = |0, 0, \ldots, 0\rangle$ with all the qubits being in the ground state. An arbitrary state $|\Psi\rangle$ of the assembly may be generated via the action of a polynomial $F(\sigma_i^+)$ in the nilpotent operators σ_i^+ on the Fock vacuum:

$$
\begin{aligned}
|\Psi\rangle &= \sum_{\{k_i\}=0,1} \psi_{k_n k_{n-1} \ldots k_1} |k_n, k_{n-1}, \ldots, k_1\rangle \quad (5) \\
&= \psi_{00\ldots0} |0, 0, \ldots 0\rangle + \psi_{10\ldots0} |1, 0, \ldots 0\rangle \\
&\quad + \psi_{01\ldots0} |0, 1, \ldots 0\rangle + \ldots + \psi_{11\ldots1} |1, 1, \ldots 1\rangle \\
&= (\psi_{00\ldots0} + \psi_{00\ldots1}\sigma_1^+ + \ldots + \psi_{01\ldots0}\sigma_{n-1}^+ + \psi_{10\ldots0}\sigma_n^+ \\
&\quad + \psi_{00\ldots11}\sigma_2^+\sigma_1^+ \ldots + \psi_{11\ldots0}\sigma_n^+\sigma_{n-1}^+ + \ldots) |O\rangle \\
&= \sum_{\{k_i\}=0,1} \psi_{k_n k_{n-1} \ldots k_1} \prod_{i=1}^n (\sigma_i^+)^{k_i} |O\rangle,
\end{aligned}
$$

Here, the subscript i enumerates the qubits, and the operator σ_i^+ creates the state $|1\rangle$ out of the state $|0\rangle$ of i-th qubit. The family of all polynomials $F(\sigma_i^+)$ forms a ring, which means that it admits addition and multiplication, but division might be ambiguous. We note that the nilpotent variables introduced here commute with one another, in contrast to anticommuting nilpotent (Grassmann) variables widely employed in quantum field theory.

It is convenient to impose a non-standard normalization condition $|\langle O |\Psi\rangle_O| = 1$, that is

$$
\begin{aligned}
F(\{\sigma_i^+\}) &= \sum_{\{k_i\}=0,1} \alpha_{k_n k_{n-1} \ldots k_1} \prod_{i=1}^n (\sigma_i^+)^{k_i} \quad (6) \\
&= \sum_{\{k_i\}=0,1} \frac{\psi_{k_n k_{n-1} \ldots k_1}}{\psi_{00\ldots0}} \prod_{i=1}^n (\sigma_i^+)^{k_i} .
\end{aligned}
$$

Such a normalization is practical, since we mainly work not with F by itself, but with its extensive counterpart – nilpotential – a nilpotent polynomial

$$f(\{\sigma_i^+\}) = \ln\left[F(\{\sigma_i^+\})\right] \tag{7}$$

$$= \sum_{\{k_i\}=0,1} \beta_{k_n k_{n-1}...k_1} \prod_{i=1}^{n} \left(\sigma_i^+\right)^{k_i} ,$$

where the coefficients $\beta_{k_n k_{n-1}...k_1}$ and $\alpha_{k_n k_{n-1}...k_1}$ can be explicitly related to each other by expanding $\ln F$ in a Taylor series around 1. Since $(\sigma_i^+)^2 = 0$, this series is a polynomial containing at most $2^n - 1$ terms.

In order to uniquely characterize entanglement, as a convenient orbit marker we take a state $|\Psi_c\rangle_{\mathcal{O}}$ lying in the orbit $\mathcal{O}$, which is the closest to the reference state $|O\rangle$ in the inner product sense, that is $|\langle O |\Psi_c\rangle_{\mathcal{O}}| = \max$. Once the state $|\Psi_c\rangle_{\mathcal{O}}$ is found, we impose the normalization condition $|\langle O |\Psi_c\rangle_{\mathcal{O}}| = 1$. The resulting canonic state $|\Psi_c\rangle_{\mathcal{O}}$ is associated with the canonic form of the polynomial $F_c(\sigma_i^+)$, which begins with a constant term equal to 1. Such a choice yields the tanglemeter – uniquely defined as a nilpotent polynomial $f_c = \ln F_c$. Both the tanglemeter and nilpotential resemble to the eikonal, which is the logarithm of the regular *semi-classical* wave function in the position representation, multiplied by $-i$. The difference is that in our case no approximation is made: f represents the logarithm of the exact state vector.

The nilpotential f and the tanglemeter f_c have several remarkable properties: *(i)* the tanglemeter provides a unique and extensive characterization of entanglement; *(ii)* a straightforward entanglement criterion can be stated in terms of the cross derivatives $\partial^2 f/\partial\sigma_i^+ \partial\sigma_j^+$: **The entanglement criterion:** *The parts A and B of a binary partition of an assembly of n qubits are unentangled iff*

$$\frac{\partial^2 f_c(\{x_i\})}{\partial x_k \partial x_m} = 0, \quad \forall k \in A, : \forall m \in B. \tag{8}$$

In other words, the subsystems A and B of the partition are disentangled iff $f_{A\cup B} = f_A(\{x_{\in A}\}) + f_B(\{x_{\in B}\})$, and no cross terms are present in the tanglemeter. Note that this criterion holds not only for the tanglemeter f_c, but for the nilpotential f as well; *(iii)* the dynamic equation of motion for f can be written explicitly and, suggestively, in the rather general case has the same form as the well-known classical Hamilton-Jacobi equation for the eikonal.

4. Number of the parameters characterizing entanglement

Consider n qubits in a generic pure state $|\Psi\rangle$ of Eq.(5), specified by 2^n complex amplitudes $\psi_{k_n k_{n-1} \ldots k_1}$, i.e. by 2^{n+1} real numbers. When we take normalization into account and disregard the global phase, there are $2^{n+1} - 2$ real parameters characterizing the assembly state. The first relevant question is: What is the maximum number of real numbers needed for the orbit identification, hence, for entanglement characterization?

As it has already been stated earlier, it is natural to require that any measure characterizing the intrinsic entanglement in the assembly state remains invariant under unitary transformations changing the state of each qubit. A generic $SU(2)$ transformation is the exponential of an element of the $su(2)$ algebra,

$$U = \exp[\mathrm{i}(\sigma^x P^x + \sigma^y P^y + \sigma^z P^z)] \,, \tag{9}$$

where σ^x, σ^y, and σ^z are Pauli matrices. It depends on the three real parameters P^x, P^y, and P^z. Therefore, for n qubits the dimension of the coset $\mathcal{H}/\mathcal{O}$, that is the number of different real parameters invariant under local unitary transformations, reads

$$D_{su} = 2^{n+1} - pn - 2 \,, \tag{10}$$

where $p = 3$ the number of the local transformation parameters for single qubit. To be precise, the counting Eq. (10) is true for $n > 2$ while the case $n = 2$ is special: in spite of the fact that $2^3 - 3 \cdot 2 - 2 = 0$, there is a nontrivial invariant of local transformations for two qubits of the form Eq.(2).

The coset dimension also depends on the symmetry group of local operations. For example, certain applications involving indirect measurements and known as SLOCC maps rely on a class of local transformations constrained only by the requirement of unit determinant. They are described by the complexification $sl(2, \mathbb{C})$ of the $su(2)$ algebra, such that the parameters (P_i^x, P_i^y, P_i^z) specifying the transformation of Eq. (9) on each qubit are now complex numbers. Therefore in Eq.(10) $p = 6$ for this case.

5. Examples: Canonic forms for two, three, and four qubits

For two qubits the result is immediate

$$f_c = \beta_{11}\sigma_2^+\sigma_1^+, \qquad F_c = 1 + \alpha_{11}\sigma_2^+\sigma_1^+ = 1 + \beta_{11}\sigma_2^+\sigma_1^+ \,, \tag{11}$$

where the constant $\alpha_{11} = \beta_{11}$ can be chosen real. For three qubits, the canonic forms of F and f also differ only by the unity term,

$$f_c = \beta_3\sigma_2^+\sigma_1^+ + \beta_5\sigma_3^+\sigma_1^+ + \beta_6\sigma_3^+\sigma_2^+ + \beta_7\sigma_3^+\sigma_2^+\sigma_1^+ = F_c - 1 \,. \tag{12}$$

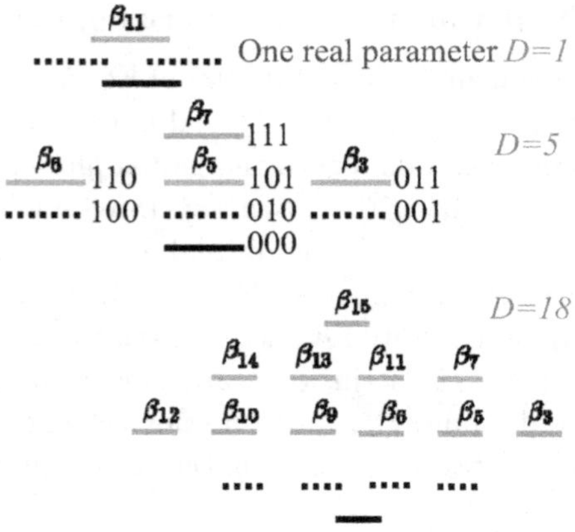

Figure 5.   Correspondence between the tanglemeter coefficients and the energy levels for two, three and four qubit cases. The states just below the maximum energy state, all have real coefficients β.

Here, we have introduced a shorter notation by considering the indices of β as binary representation of decimal numbers, $011 \to 3$, *etc.* One can make use of the fact that the variables σ_i^+ are defined up to phase factors, and set β_3, β_5, and β_6 real.

The tanglemeter for four qubits reads

$$
\begin{aligned}
f_c = {} & \beta_3 \sigma_2^+ \sigma_1^+ + \beta_5 \sigma_3^+ \sigma_1^+ + \beta_9 \sigma_4^+ \sigma_1^+ + \beta_6 \sigma_3^+ \sigma_2^+ + \beta_{10} \sigma_4^+ \sigma_2^+ \\
& + \beta_{12} \sigma_4^+ \sigma_3^+ + \beta_7 \sigma_3^+ \sigma_2^+ \sigma_1^+ + \beta_{13} \sigma_4^+ \sigma_3^+ \sigma_1^+ + \beta_{11} \sigma_4^+ \sigma_2^+ \sigma_1^+ \\
& + \beta_{14} \sigma_4^+ \sigma_3^+ \sigma_2^+ + \beta_{15} \sigma_4^+ \sigma_3^+ \sigma_2^+ \sigma_1^+,
\end{aligned}
\tag{13}
$$

while the coefficients α_i of the polynomial F_c differ from β_i only at the last position

$$
\alpha_{15} = \beta_{15} + \beta_3 \beta_{12} + \beta_5 \beta_{10} + \beta_9 \beta_6.
\tag{14}
$$

One may note that the sums of the indices of the factors in this expression are equal. The latter is a general feature for the relationship among the coefficients α and β: the coefficients α are given by sums of terms, each of which contains a product of the coefficients β where the sum of the indices equals the index of α.

The next example corresponds to the first non-trivial case of local SL symmetry – the three-qubit case. The *sl*-tanglemeter for a generic three-qubit state reads

$$
f_C = \sigma_3^+ \sigma_2^+ \sigma_1^+,
\tag{15}
$$

where the coefficient is set to unity by the scale freedom in the definition of the nilpotent variables. The corresponding wave function F_C is nothing but the GHZ state. This shows again that all generic states belong to the same *sl*-orbit, which includes this state. There are, however, also three distinct singular classes of entangled states of measure zero whose tanglemeters do not involve the product $\sigma_1^+ \sigma_2^+ \sigma_3^+$ and have one of the following forms,

$$f_C = \sigma_2^+ \sigma_1^+ + \sigma_3^+ \sigma_1^+,$$

$$f_C = \sigma_3^+ \sigma_1^+ + \sigma_3^+ \sigma_2^+,$$

$$f_C = \sigma_2^+ \sigma_1^+ + \sigma_3^+ \sigma_2^+.$$

In this classification, we have only taken into account the states whose tanglemeters involve all three σ_i^+ such that no qubit is completely disentangled from the others. For a generic four-qubit state one finds the *sl*-tanglemeter

$$f_C = \beta_3 \sigma_2^+ \sigma_1^+ + \beta_5 \sigma_3^+ \sigma_1^+ + \beta_9 \sigma_4^+ \sigma_1^+ + \beta_6 \sigma_3^+ \sigma_2^+ \tag{16}$$
$$+ \beta_{10} \sigma_4^+ \sigma_2^+ + \beta_{12} \sigma_4^+ \sigma_3^+ + \beta_{15} \sigma_4^+ \sigma_3^+ \sigma_2^+ \sigma_1^+,$$

6. Quantum dynamics in terms of the nilpotent polynomials

One can describe dynamics of an assembly state vector as evolution of the corresponding polynomials F and f. As an example illustrating the usefulness of the time dependent tanglemeter, we present the evolution of the tanglemeter coefficients that emerge from implementation of the Grover's search algorithm.[1] It turns out that the only tanglemeter coefficient different from zero stands in front of the product of the creation operators $\sigma_{i \in R}^+$ corresponding to the qubits that are at the state $|1\rangle$ at the end of the calculation

$$f_c = \beta_G(t) \prod_{i \in R} \sigma_i^+.$$

More specifically, the assembly of qubits that forms the quantum register, evolves directly from the initial vacuum state to the final product state $\prod_{i \in R} \sigma_i^+ |O\rangle$, which corresponds to the solution of the search problem: the qubits that belong to a variety R are in the upper state while the rest of the qubits are in the lower state. The parameter $\beta_G(t)$ linearly increases from zero at the beginning, to its maximum value 1 in the middle of the process and then linearly returns back to zero at the end of the calculation, as the

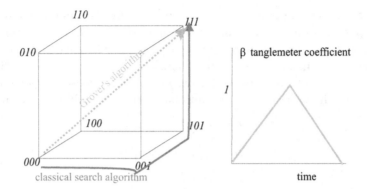

Figure 6. Execution of the Grover's search algorithm moves the quantum register directly from the initial product state to the final product state through a set of entangled states. States of the set are entangled in such a way that only one of the tanglemeter coefficient differs from zero. This coefficient is in front of the term $\prod_{i \in R} \sigma_i^+$, where R is the set of qubits that will be in the excited states at the end of the calculation. It linearly increases with time from zero value at the beginning, takes the maximum value 1 at the middle of the calculation, and linearly returns back to zero value.

register returns back to a factorizable state. Let us now turn to the general situation and derive a dynamic equation for the nilpotential of a quantum assembly. To this end, we focus at infinitesimal transformations, and find the equations of motion describing the dynamics of the nilpotential under continuous local and gate operations. It turns out that for an important class of Hamiltonians supporting universal quantum computation, the dynamic equation for the nilpotential acquires a well-known Hamilton-Jacobi form.

Consider an infinitesimal transformations acting on the state vector $|\Psi\rangle = F(\{\sigma_i^+\})|O\rangle$ and yielding a transformed state $F'(\{\sigma_i^+\})|O\rangle$. One can formalize the rules allowing one to obtain F' from F. Bearing in mind that $\sigma^-|0\rangle = 0$ and $\sigma^z|0\rangle = -|0\rangle$, one can represent the action of $\sigma_i^+, \sigma_i^z, \sigma_i^-$ operators on the state vector as appropriate differential operations acting on the polynomials $F(\sigma_i^+)$. The application of the operator σ_i^+ is straightforward – it is a direct multiplication: this operation eliminates the terms that were proportional to σ_i^+ prior to the multiplication. The application of σ_i^- is a kind of inverse: it can be considered as a derivative with respect to the variable σ_i^+, which eliminates the terms independent of σ_i^+ and makes the terms linear in σ_i^+ independent of this variable. Finally, the application of σ_i^z changes the signs of the terms independent of σ_i^+, and leaves intact terms linear in σ_i^+. These actions are summarized by the following formulae

$$\sigma_i^+ F \rightarrow \sigma_i^+ F\,,$$

$$\sigma_i^- F \rightarrow \frac{\partial F}{\partial \sigma_i^+}\,, \qquad (17)$$

$$\sigma_i^z F \rightarrow -F + 2\sigma_i^+ \frac{\partial F}{\partial \sigma_i^+}\,,$$

while each unitary operation $U\left(\sigma_i^x, \sigma_i^y, \sigma_i^z\right)$ acting on the state vector can be represented by a differential operator $U\left(\sigma_i^+, \frac{\partial}{\partial \sigma_i^+}, 2\sigma_i^+ \frac{\partial}{\partial \sigma_i^+} - 1\right)$ acting on the polynomial

$$F' = U\left(\left\{\sigma_i^+, \frac{\partial}{\partial \sigma_i^+}, 2\sigma_i^+ \frac{\partial}{\partial \sigma_i^+} - 1\right\}\right) F\,, \qquad (18)$$

while for the nilpotential one immediately finds

$$f' = \log\left(U\left(\left\{\sigma_i^+, \frac{\partial}{\partial \sigma_i^+}, 2\sigma_i^+ \frac{\partial}{\partial \sigma_i^+} - 1\right\}\right) e^f\right)\,. \qquad (19)$$

Note that a generic transformation Eq. (19) of an initially canonic polynomial does not necessarily results in another canonic polynomial.

Consider now an infinitesimal unitary transformation $U = 1 - \mathrm{i}\, \mathrm{d}t\, H$ which is not necessarily local. The increment Δf of the nilpotential f suggested by the Eq. (19) reads

$$\Delta f = \log\left(U e^f\right) - \log\left(e^f\right) = \log\left(1 - \mathrm{i}\, \mathrm{d}t\, e^{-f} H e^f\right)\,. \qquad (20)$$

This yields the following dynamic equation for f:

$$\mathrm{i}\frac{\partial f}{\partial t} = e^{-f} H e^f\,. \qquad (21)$$

For the particular case of the local Hamiltonian

$$H_L = \sum_i P_i^x(t)\sigma_i^x + P_i^y(t)\sigma_i^y$$

$$= \sum_i P_i^-(t)\sigma_i^+ + P_i^+(t)\sigma_i^-\,, \qquad (22)$$

and the binary interaction

$$H_B = \sum_{i,j} G_{ij}(t)\sigma_i^+ \sigma_j^-\,, \qquad (23)$$

that together ensure the universal evolution of the qubit assembly, one finds the explicit equation

$$
\mathrm{i}\frac{\partial f}{\partial t} = \sum_i \left[P_i^-(t)\sigma_i^+ + P_i^+(t)\frac{\partial f}{\partial \sigma_i^+}\left(1 - \sigma_i^+\frac{\partial f}{\partial \sigma_i^+}\right)\right] \qquad (24)
$$
$$
+ \sum_{i\neq j} G_{ij}(t)\sigma_j^+\frac{\partial f}{\partial \sigma_i^+}\left(1 - \sigma_i^+\frac{\partial f}{\partial \sigma_i^+}\right).
$$

Note that Eq. (24) for nilpotential formally resembles the Hamilton-Jacobi equation for the mechanical action of classical systems with the Hamiltonian

$$
H = \sum_i \left[P_i^-(t)x_i + P_i^+(t)p_i\left(1 - x_i p_i\right)\right]
$$
$$
+ \sum_{i\neq j} G_{ij}(t)\left[x_j p_i\left(1 - x_i p_i\right)\right]. \qquad (25)
$$

where $p_i = \partial f/\partial \sigma_i^+$ plays a role of the momentum, while $x_i = \sigma_i^+$ are the coordinates. Comparing with the conventional classical Hamilton-Jacobi equation, the only essential difference is the factor i multiplying the time derivative and the presence of complex parameters that can be interpreted as time-dependent forces and masses.

7. Entanglement beyond qubits

The nilpotent polynomials approach may be extended to describe situations more general than assemblies of qubits. Canonic states of such assemblies have already been discussed, and now the question is: what one has to employ in the place of the Pauli operators σ^+ in order to construct an extensive nilpotent characteristic? The construction of nilpotent polynomials for such systems is based on the so-called Cartan-Weyl decomposition. Here we illustrate this for qutrits corresponding to local $SU(3)$ symmetry group. This are eight generators represented by the Gell-Mann λ^a matrices that now play role of the Pauli matrices. In particular two matrices

$$
\lambda^3 = \begin{pmatrix} 1 & 0 & 0 \\ 0 & -1 & 0 \\ 0 & 0 & 0 \end{pmatrix}, \lambda^8 = \frac{1}{\sqrt{3}}\begin{pmatrix} 1 & 0 & 0 \\ 0 & 1 & 0 \\ 0 & 0 & -2 \end{pmatrix} \qquad (26)
$$

play role of σ^z and compose so-called Cartan subalgebra L^z. Subalgebra of rising operators L^+ is comprised of 3 elements,

$$s^+ = \frac{\lambda^1 + i\lambda^2}{2} = \begin{pmatrix} 0 & 1 & 0 \\ 0 & 0 & 0 \\ 0 & 0 & 0 \end{pmatrix}, \tag{27}$$

$$u^+ = \frac{\lambda^4 + i\lambda^5}{2} = \begin{pmatrix} 0 & 0 & 1 \\ 0 & 0 & 0 \\ 0 & 0 & 0 \end{pmatrix}, \tag{28}$$

$$t^+ = \frac{\lambda^6 + i\lambda^7}{2} = \begin{pmatrix} 0 & 0 & 0 \\ 0 & 0 & 1 \\ 0 & 0 & 0 \end{pmatrix}, \tag{29}$$

and only two of them, the commuting pair u^+ and t^+, are required for construction of the nilpotent polynomials.

A generic pure state of a qutrit may be represented as

$$|\psi\rangle = \psi_0|0\rangle + \psi_1|1\rangle + \psi_2|2\rangle = \left(\psi_0 + \psi_1 t^+ + \psi_2 u^+\right)|0\rangle, \tag{30}$$

with

$$|0\rangle \equiv \begin{pmatrix} 0 \\ 0 \\ 1 \end{pmatrix}, \; |1\rangle \equiv \begin{pmatrix} 0 \\ 1 \\ 0 \end{pmatrix}, \; |2\rangle \equiv \begin{pmatrix} 1 \\ 0 \\ 0 \end{pmatrix}, \tag{31}$$

which generalizes a similar representation for the qubit pure states. Note that even for larger d, such a set of $d-1 = r$ commuting nilpotent operators always exists, allowing one to express a generic qudit state as a first-order polynomial of commuting nilpotent variables.

The state of n qutrits can then be written as

$$\begin{aligned} |\Psi\rangle &= \sum_{\nu_i, \eta_i = 0,1} \psi_{2\nu_n + \eta_n \ldots 2\nu_1 + \eta_1} \prod_{i=1}^{n} (u_i^+)^{\nu_i} (t_i^+)^{\eta_i} |O\rangle \\ &= F(u_i^+, t_i^+)|O\rangle, \end{aligned} \tag{32}$$

where $|O\rangle = |0, \ldots, 0\rangle$ now denotes the vacuum state of qutrits. For the canonic state the polynomial $F(u_i^+, t_i^+)$ does not have terms linear in u_i^+ and t_i^+ and starts with the bilinear terms. Cubic terms, in turn, do not contain terms linear in u_i^+, which follows from the requirement of the maximum overlap with the state $|1\rangle$ as it is illustrated in Fig.7. Total number of the parameters characterizing the entanglement amounts to

$$D_{su} = 2 \cdot 3^n - 8n - 2, \tag{33}$$

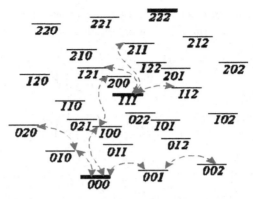

Figure 7.   The states $|211\rangle, |121\rangle$, and $|112\rangle$ corresponding to qubic terms linear in u_i^+ are depopulated by the requirement of the maximum overlap with the state $|1\rangle$

which is valid for $n \geq 3$

For two qutrits one obtains

$$F_c(u_2^+, t_2^+, u_1^+, t_1^+) = 1 + \alpha_{11}t_2^+ t_1^+ + \alpha_{22}u_2^+ u_1^+ \qquad (34)$$

$$f_c(u_2^+, t_2^+, u_1^+, t_1^+) = \alpha_{11}t_2^+ t_1^+ + \alpha_{22}u_2^+ u_1^+. \qquad (35)$$

These polynomials depend only on two complex parameters α_{ii}, which can be set real and positive by a local phase transformation,

$$\exp\left(i\gamma_2\lambda_2^3 + i\delta_2\lambda_2^8 + i\gamma_1\lambda_1^3 + i\delta_1\lambda_1^8\right). \qquad (36)$$

Thus, two real parameters are sufficient for characterizing entanglement between two qutrits, which is consistent with the result obtained by a straightforward application of the bipartite Schmidt decomposition.

In order to better understand the pattern, consider two more examples: an assembly of three qutrits, and of two qutrits and a qubit. In the first case, the canonic form is

$$\begin{aligned}
F_c = 1 &+ \alpha_{011}t_2^+ t_1^+ + \alpha_{012}t_2^+ u_1^+ + \alpha_{021}u_2^+ t_1^+ + \alpha_{022}u_2^+ u_1^+ + \alpha_{101}t_3^+ t_1^+ \\
&+ \alpha_{102}t_3^+ u_1^+ + \alpha_{201}u_3^+ t_1^+ + \alpha_{202}u_3^+ u_1^+ + \alpha_{110}t_3^+ t_2^+ + \alpha_{120}t_3^+ u_2^+ \\
&+ \alpha_{210}u_3^+ t_2^+ + \alpha_{220}u_3^+ u_2^+ + \alpha_{111}t_3^+ t_2^+ t_1^+ + \alpha_{122}t_3^+ u_2^+ u_1^+ \qquad (37) \\
&+ \alpha_{212}u_3^+ t_2^+ u_1^+ + \alpha_{221}u_3^+ u_2^+ t_1^+ + \alpha_{222}u_3^+ u_2^+ u_1^+ ,
\end{aligned}$$

whereas for two qutrits and a qubit (labeled by the index 3), it looks as follows

$$F_c = 1 + \alpha_4 t_2^+ t_1^+ + \alpha_5 t_2^+ u_1^+ \alpha_7 u_2^+ t_1^+ + \alpha_8 u_2^+ u_1^+ + \alpha_{10} t_3^+ t_1^+$$
$$+ \alpha_{11} \sigma_3^+ u_1^+ + \alpha_{12} \sigma_3^+ t_2^+ + \alpha_{15} \sigma_3^+ u_2^+ + \alpha_{13} \sigma_3^+ t_2^+ t_1^+ \qquad (38)$$
$$+ \alpha_{17} \sigma_3^+ u_2^+ u_1^+ .$$

8. Generalized entanglement

The nilpotent polynomials formalism allows us to consider a special case of generalized entanglement where due to some constraint, the algebra of local transformations does not include all possible transformations that exist for an element of the dimension d. In this context, of special interest are spin-1 systems, namely three-level systems restricted to evolve under the action of spin operators living in the $so(3) \equiv su(2)$ subalgebra of $su(3)$. In such an assembly, entanglement may exist even within a single element, which still can be characterized via the nilpotent polynomials. For the purpose, one has to invoke the nilpotent variables of higher order whose squares do not vanish and only some higher powers do. With the help of these nilpotent variables one can also consider entanglement among different elements of the assembly.

Though a three-level system corresponds to a full $su(3)$ algebra, we consider it here as a spin-1 system that is, concentrate on a situation where the physical observables are restricted to the subalgebra $su(2)$ of spin operators, $S_\pm = S_x \pm iS_y$ and S_z. Note that the latter are equivalent to the $u^\pm + t^\mp$, and λ^3 generators of $su(3)$, respectively. The spin states are characterized by the eigenvalues of

$$\lambda^3 = \begin{pmatrix} 1 & 0 & 0 \\ 0 & 0 & 0 \\ 0 & 0 & -1 \end{pmatrix},$$

and he spin-down state $|-1\rangle$ is chosen as the reference state. Here, in order to make use of the standard spin-1 vector notations, we have interchanged positions of the second and the third lines with respect of Eq.(26). The operator

$$S_+ = u^+ + t^- = \begin{pmatrix} 0 & 1 & 0 \\ 0 & 0 & 1 \\ 0 & 0 & 0 \end{pmatrix}$$

is the only element of the nilpotent subalgebra $su(2) \subset su(3)$, which has therefore to be chosen as the nilpotent variable. It is of the order of two since $S_+^2 \neq 0$, and only $S_+^3 = 0$.

Now we show that even for a single spin, the state $|0\rangle$ is generalized entangled with respect to $SU(2)$. Indeed, by the unitary matrix $e^{-i\pi S_z/\sqrt{2}}$ the state $|0\rangle$ can be transformed to the state with the maximum vacuum population $|\psi_C\rangle = (|-1\rangle + |1\rangle)/\sqrt{2}$, which evidently differs from the reference state $|-1\rangle$. The corresponding canonic state normalized to unit reference state amplitude reads

$$|\Psi_c\rangle = (1 + S_+^2)|-1\rangle, \tag{39}$$

hence $f_c = S_+^2$ and $F_c = 1 + S_+^2$. The presence of the quadratic term is the signature of generalized entanglement.

One way to understand the meaning of this "self-entanglement" in the state $|0\rangle$ is to see it as a consequence of the fact that the operators $S_{\pm,z}$ of $su(2)$ cannot lift the degeneracy of the two eigenstates of the operator λ_8, which together with $\lambda_3 = S_z$ labels the states in the unrestricted $su(3)$ algebra. In other words, within the group of restricted local transformations, the transition from the state $|0\rangle$ can only access the state $(|-1\rangle + |1\rangle)/\sqrt{2}$ but not $(|-1\rangle - |1\rangle)/\sqrt{2}$, hence the amplitudes of the reference state $|-1\rangle$ and that of the state $|1\rangle$ are fully correlated. One will not be too much surprised with existence of the self-entanglement after paying attention to the fact that here as well as earlier the subgroup of allowed local transformations of the element is also smaller than the entire group of possible transformation of this element (which is strictly speaking also local, although not completely allowed).

For a generic state, the canonic nilpotent polynomial and the tanglemeter take the form

$$F_c(S_+) = 1 + \frac{\alpha_s}{2}S_+^2$$

$$f_c(S_+) = \frac{\alpha_s}{2}S_+^2, \tag{40}$$

respectively, where the phase of α'_s can be set to 0, and the factor $\frac{1}{2}$ is introduced for convenience. Therefore, generalized entanglement of a single spin-1 is characterized by a single real parameter.

Consider now entanglement between two spin-1, which we describe as two three-level systems subject to the action of $SU(2) \oplus SU(2)$ local operations. By analogy to the single spin-1 case, we chose $S_{1,2}$ with the implicit subscripts $+$ as nilpotent variables and the state $|-1,-1\rangle$ as the reference. In the canonic form maximizing population of the reference state, one obtains

$$F_c(S_1, S_2) = 1 + \alpha_{s;1}\frac{S_1^2}{2} + \alpha_{s;2}\frac{S_2^2}{2} + \alpha_{s,s}\frac{S_2^2}{2}\frac{S_1^2}{2} \tag{41}$$

$$+ \alpha_{t,s}\frac{S_2^2}{2}S_1 + \alpha_{st}S_2\frac{S_1^2}{2} + \alpha_{t,t}S_1 S_2 ,$$

$$f_c(S_1, S_2) = \beta_{s;1}\frac{S_1^2}{2} + \beta_{t,t}S_1 S_2 + \beta_{st}\frac{S_1^2}{2}S_2 + \beta_{s;2}\frac{S_2^2}{2}$$

$$+ \beta_{t,s}S_1\frac{S_2^2}{2} + \beta_{s,s}\frac{S_2^2}{2}\frac{S_1^2}{2} ,$$

where $\beta_{s,s} = \alpha_{s,s} - 2\alpha_{t,t}^2$, and all other $\beta = \alpha$. As before, by exploiting the freedom of phase transformations on the nilpotent variables, the parameters $\alpha_{s;1}$ and $\alpha_{s;2}$ (or $\beta_{s;1}$ and $\beta_{s;2}$) characterizing generalized entanglement within each of the three-level systems can be set real, and we are left with four complex numbers $\alpha_{s,s}$, $\alpha_{t,s}$, α_{st}, and $\alpha_{t,t}$ characterizing the generalized inter-spin entanglement.

9. A step toward mixed states of an assembly

Thus far we have addressed only pure states of an assembly, where all manipulations with the corresponding state vector were formulated in group theoretical terms. The relevant algebraic technique relies on two basic operations in the exponent – the summation and the algebraic product. Addressing mixed states implies working with the density operator, which is a statistically weighted sum over the binary tensor products of possible state vectors and their Hermitian conjugates. Transformation of the density operators suggested by the group theory does not offer full access to manipulation of the statistical weights. In particular, the local unitary group does not change them at all, and only the group of general invertible local transformations is capable of changing these quantities. Trying to characterize entanglement for the mixed states, we therefore go beyond the domain where algebraic approach looks as a promising tool. The reason for this is in the fact that apart from the two standard operations in the exponent, the summation and the algebraic binary product, a third operation namely the summation of the exponents is indispensable for the state mixing. Moreover, the very definition of entanglement for mixed states essentially relies on this operation, – a state is considered as not entangled when can be represented as a sum of statistically weighted product states.

Description of the entanglement hence encounters a difficulty,– the same density operator can result from various statistical mixtures of different states. In particular, equally weighted statistical mixture of maximum entangled pure states that belong to the same orbit may and does yield a density operator identical to that of a product states mixture, and hence has to be considered unentangled. If we forget for a while this difficulty and adopt an alternative criterion of so-called distillable entanglement[2] discussed below, the classification of the mixed state entanglement can rely on a conceptually transparent idea: bra$\langle|$ and ket $|\rangle$ vectors of the same element are considered as two independent ket vectors belonging to two distinct but tween elements. This way one arrives at a pure state of an assembly with twice more elements, which can be described by nilpotent polynomial as earlier, with the only constraint that the Hermitian symmetry of the density operator must be respected. In such a framework, entanglement among the ket state of an original element and a twin bra state of the same or another element results in the state mixing.

Let us consider an assembly of qubits and denote by σ_i^+ and $\overline{\sigma}_i^+$ the rising operator for i-th qubit and for it tween, respectively. One can immediately identify three limiting cases: (i) the tanglemeter has no cross terms containing σ_i^+ and $\overline{\sigma}_j^+$ at once, which means that the state is pure; (ii) the tanglemeter has no cross terms among the operators (no matter with or without bar) with indexes $i \in A$ and $j \in \overline{A}$, which means that the density operator is a tensor product of the density operators of the group of qubits A and that of the rest of the qubits $\overline{A}$; and (iii) the tanglemeter contains only the second order cross terms $\sigma_i^+ \overline{\sigma}_j^+$, which means that this is a statistical mixture of product states, as one can infer by applying the Hubbard-Stratanovich separation of the operator products in the exponent. The condition (iii) is sufficient for the mixed state to be unentangled. However, one cannot be yet be sure, whether or not it is necessary.

Let us make one step further in the consideration by focussing at mixed states of two qubits. By the analogy to Eq.(14) for the tanglemeter of a 4-qubit system in the pure state, one can write

$$
\begin{aligned}
f_c = {} & \gamma_{0,3}\sigma_2^+\sigma_1^+ + \gamma_{3,0}\overline{\sigma}_2^+\overline{\sigma}_1^+ + \gamma_{3,3}\overline{\sigma}_2^+\overline{\sigma}_1^+\sigma_2^+\sigma_1^+ \\
& + \gamma_{1,1}\overline{\sigma}_1^+\sigma_1^+ + \gamma_{2,1}\overline{\sigma}_2^+\sigma_1^+ + \gamma_{1,2}\overline{\sigma}_1^+\sigma_2^+ + \gamma_{2,2}\overline{\sigma}_2^+\sigma_2^+ \\
& + \gamma_{1,3}\overline{\sigma}_1^+\sigma_2^+\sigma_1^+ + \gamma_{3,1}\overline{\sigma}_2^+\overline{\sigma}_1^+\sigma_1^+ \\
& + \gamma_{2,3}\overline{\sigma}_2^+\sigma_2^+\sigma_1^+ + \gamma_{3,2}\overline{\sigma}_2^+\overline{\sigma}_1^+\sigma_2^+
\end{aligned}
\tag{42}
$$

for the mixed state tanglemeter, where the hermiticity of the density operator is taken into account by the fact that coefficients $\gamma_{i,j}$ comprise a Hermitian matrix.

With the help of Eq.(42) after straightforward calculations one finds the density operator of two qubits:

$$
\rho(\gamma) = \begin{pmatrix} |\gamma_{0,3}|^2 + \gamma_{22}\gamma_{1,1} + |\gamma_{1,2}|^2 + \gamma_{3,3} & \gamma_{2,3} & \gamma_{1,3} & \gamma_{0,3} \\ \gamma_{3,2} & \gamma_{2,2} & \gamma_{1,2} & 0 \\ \gamma_{3,1} & \gamma_{2,1} & \gamma_{1,1} & 0 \\ \gamma_{3,0} & 0 & 0 & 1 \end{pmatrix}
$$

normalized by the condition $\rho_{0,0} = 1$. Now one can calculate the difference C between the largest eigenvalue and the sum of all the rest eigenvalues of the concurrence tensor $\sqrt{\rho(\gamma) \cdot \epsilon_1 \otimes \epsilon_2 \cdot \rho^*(\gamma) \cdot \epsilon_1 \otimes \epsilon_2}$, where dot $\cdot$ stands for the matrix multiplication in contrast to the tensor product notation $\otimes$, and ϵ_i is the antisymmetric tensor for qubit i, which has been already employed earlier in Sec.I when we were constructing the invariants of local transformations for pure states of assemblies. For the non-positive difference C, the qubits are proven to be not entangled. Thus we have reformulated the well-known entanglement criterion for mixed states of two qubits in terms of the nilpotent polynomials.

There exists an alternative to the concurrence, so-called criterion of the distillable entanglement, suggested by the requirement that an assembly in a mixed state can be set to a pure entangled state by a sequence of local sl transformations implied by indirect local measurements. For a two qubit mixed state given by the density operator ρ, one can directly check whether or not this criterion holds. For the purpose one has to notice that for such a final state, the tanglemeter matrix elements $\gamma_{i,j}$ vanish for all pairs $\{i,j\}$ except of $\{0,3\}$ and $\{3,0\}$, that is the density operator satisfying the criterion of entanglement in general case reads

$$
\rho_{En} = e^{-i\mathbb{P}^*} \begin{pmatrix} |\gamma_{0,3}|^2 & 0 & 0 & \gamma_{0,3} \\ 0 & 0 & 0 & 0 \\ 0 & 0 & 0 & 0 \\ \gamma_{3,0} & 0 & 0 & 1 \end{pmatrix} e^{i\mathbb{P}}, \tag{43}
$$

where

$$
\mathbb{P} = \sum_{i=1,2} (\sigma_i^x P_i^x + \sigma_i^y P_i^y + \sigma_i^z P_i^z). \tag{44}
$$

The seven complex numbers $\gamma_{0,3}$ and $P_{1;2}^{x;y;z}$ determine the family of the density operators satisfying this criterion. They are uniquely defined by the requirement $\rho = \rho_{En}$ imposed on the density operator of the assembly, when this equation has a solution. Answering the question whether or not this formalism is constructive for larger assemblies is a matter of future research, although already for three qubit assembly, by a simple counting of parameters, one finds that the generic mixed state cannot be reduced to a pure multipartite entangled states by local SL transformations. The latter approach in the case of large assemblies could also rely on the dynamic equation similar to Eq.(24) for the nilpotential, which now takes the form

$$
\begin{aligned}
i\frac{\partial f}{\partial t} = \sum_i & \left[P_i(t)\sigma_i^+ + P_i^*(t)\frac{\partial f}{\partial \sigma_i^+}\left(1 - \sigma_i^+ \frac{\partial f}{\partial \sigma_i^+}\right) \right. \\
& \left. - P_i^*(t)\overline{\sigma}_i^+ - P_i(t)\frac{\partial f}{\partial \overline{\sigma}_i^+}\left(1 - \overline{\sigma}_i^+ \frac{\partial f}{\partial \overline{\sigma}_i^+}\right) \right],
\end{aligned}
\tag{45}
$$

while the question whether or not a mixed state is distillable-entangled now can be formulated as a possibility to choose a time dependent complex-valued control vector $P_i(t)$, such that at the end of the evolution suggested by Eq.(45), the nilpotential has no nonlinear cross terms containing both $\overline{\sigma}_i^+$ and σ_j^+ at once, but does have them for $\overline{\sigma}_i^+$ and σ_i^+ separately. This question adopts a form of control problem, when one writes Eq.(45) explicitly, as a nonlinear equation for the nilpotential coefficients.

10. Conclusion

We hope that we have convinced readers that the nilpotent polynomial technique is a convenient and adequate language for entanglement characterization at least for the pure states of assemblies composed by elements with finite numbers of eigenstates. It also can be adequately extended to the well-studied case of 2-qubit assembly in mixed states, and it promises certain advantages for yet not completely explored cases of larger assemblies in mixed states.

Bibliography

1. L. K. Grover, Am. Journal of Physics, **69**, 769 (2001).
2. V. Vedral, M. B. Plenio, M. A. Rippin, and P. L. Knight, Phys. Rev. Lett. **78**, 2275 (1997); E. M. Rains, Phys. Rev. A **60**, 173 (1999).

Sudden death and sudden birth of entanglement

Ryszard Tanaś

Nonlinear Optics Division, Faculty of Physics, Adam Mickiewicz University,
61-614 Poznń, Poland
E-mail: tanas@kielich.amu.edu.pl

We compare the evolution of entanglement for a system of two two-level atoms interacting with (i) a common reservoir being in the vacuum state and (ii) a common thermal reservoir at non-zero temperature. The Markovian master equation is used to describe the evolution of entanglement which is quantified by concurrence. Phenomena of sudden death, revival, and sudden birth of entanglement are discussed. It is shown that when the atoms behave collectively and the reservoir is the vacuum, the entanglement evolution exhibits quite a rich structure. For thermal reservoir with non-zero mean number of photons this structure is gradually degraded as the temperature of the reservoir increases. For thermal reservoir the phenomenon of sudden death of entanglement becomes a standard behavior, and no asymptotic evolution can be observed. As the temperature of the reservoir increases the sudden birth of entanglement, which is a signature of collective behavior of the atoms in the vacuum, gradually disappears. The results are illustrated graphically for a number of initial states of the two-atom system.

Keywords: two-atom system, entanglement, sudden death, sudden birth.

1. Introduction

Entanglement is the key property distinguishing quantum and classical worlds. It is of fundamental importance and a necessary resource for various quantum algorithms.[1] Entanglement is a very fragile phenomenon, and it is quickly deteriorated when the quantum system interacts with the environment, so, knowing the evolution of entanglement of a quantum system in a dissipative environment is of vital importance for the quantum information processing. The time evolution of entanglement for a system of two qubits, or two two-level atoms, has been widely studied in recent years.[2-19] For more information and extensive literature on the subject see the review article, Ref. 20. A lot of discussion has been devoted to the problem of disentanglement of the two-qubit system in a finite time, despite the fact that all the matrix elements of the two-atom system decay only asymptotically, *i.e.*, when time goes to infinity. Yu and Eberly[12,21,22] coined the name *entanglement sudden death* to the process of finite-time disentanglement. Entanglement sudden death has recently been confirmed experimentally.[23] Another problem related to entanglement evolution that attracted attention is the evolution of the entangled qubits interacting with the non-Markovian reservoirs.[24-26] It has also been shown that squeezed reservoir leads to the steady-state entanglement[27] and revivals of entanglement.[28]

To describe quantitatively entanglement evolution, it is usually assumed that the two atoms are independent, each of them is embedded in its own reservoir, and they are prepared initially in an entangled state, pure or mixed, and the time evolution of entanglement quantified by the values of concurrence[29] or negativity[30,31] is studied.

If the two atoms are separated by a distance comparable to the wavelength of light emitted by the atom, or smaller, and if both atoms are coupled to a common reservoir, the entanglement evolution becomes much more complex and interesting, exhibiting not only entanglement sudden death or asymptotic decay, but entanglement can also be created during the evolution,[3,7,32] or one can observe *revival of the entanglement*[15] as well as *entanglement sudden birth*.[17,33] Entanglement sudden death and sudden birth has recently been discussed for the two atoms interacting with a common structured (non-Markovian) reservoir, within the Dicke model.[34] Experimental conditions for realization of the collective Dicke model have been studied.[35] It has also been shown[36,37] that for separate reservoirs at finite temperatures, entanglement always disappears at finite time, which means that there is always entanglement sudden death when the reservoir has finite temperature.

In this paper we compare the evolution of entanglement, measured by concurrence, for a system of two two-level atoms interacting with a common reservoir at zero temperature as well as at finite temperature. The evolution of the system is described by the Markovian master equation introduced by Lehmberg[38] and Agarwal,[39] taking into account the cooperative behavior of the atoms. We discuss the role of the cooperative behavior of the two atoms in appearance of new effects in entanglement evolution when the reservoir is the vacuum, and their disappearance when the reservoir becomes thermal field with non-zero temperature. The results are illustrated graphically to visualize the differences.

2. Master equation

We consider a system of two two-level atoms with ground states $|g_i\rangle$ and excited states $|e_i\rangle$ $(i = 1, 2)$ connected by dipole transition moments $\boldsymbol{\mu}_i$. The atoms are located at fixed positions r_1 and r_2 and coupled to all modes of the electromagnetic field, which we assume to be in a thermal state.

The reduced two-atom density matrix evolves in time according to the Markovian master equation given by[38–40]

$$
\frac{\partial \rho}{\partial t} = -i \sum_{i=1}^{2} \omega_i \left[S_i^z, \rho \right] - i \sum_{i \neq j}^{2} \Omega_{ij} \left[S_i^+ S_j^-, \rho \right]
$$

$$
- \frac{1}{2} \sum_{i,j=1}^{2} \Gamma_{ij}(1 + N) \left(\rho S_i^+ S_j^- + S_i^+ S_j^- \rho - 2 S_j^- \rho S_i^+ \right)
$$

$$
- \frac{1}{2} \sum_{i,j=1}^{2} \Gamma_{ij} N \left(\rho S_i^- S_j^+ + S_i^- S_j^+ \rho - 2 S_j^+ \rho S_i^- \right), \tag{1}
$$

where S_i^+ (S_i^-) are the raising (lowering) operators, and S_i^z is the energy operator of the ith atom, $\Gamma_{ii} \equiv \Gamma$ are the spontaneous decay rates, and N is the mean number of photons of the reservoir. We assume that the two atoms are identical. The parameters Γ_{ij} and Ω_{ij} ($i \neq j$) depend on the distance between the atoms and describe the collective damping and the dipole-dipole interaction defined, respectively, by

$$
\Gamma_{ij} = \frac{3}{2} \Gamma \left(\frac{\sin k r_{ij}}{k r_{ij}} + \frac{\cos k r_{ij}}{(k r_{ij})^2} - \frac{\sin k r_{ij}}{(k r_{ij})^3} \right), \tag{2}
$$

and

$$
\Omega_{ij} = \frac{3}{4} \Gamma \left(-\frac{\cos k r_{ij}}{k r_{ij}} + \frac{\sin k r_{ij}}{(k r_{ij})^2} + \frac{\cos k r_{ij}}{(k r_{ij})^3} \right), \tag{3}
$$

where $k = \omega_0/c$, and r_{ij} is the distance between the atoms. Here, we assume, with no loss of generality, that the atomic dipole moments are parallel to each other and are polarized in the direction perpendicular to the interatomic axis.

The density operator of the system can be represented in a complete set of basis states spanned by four product states (standard basis)

$$
|1\rangle = |g_1\rangle \otimes |g_2\rangle, \quad |2\rangle = |g_1\rangle \otimes |e_2\rangle,
$$
$$
|3\rangle = |e_1\rangle \otimes |g_2\rangle, \quad |4\rangle = |e_1\rangle \otimes |e_2\rangle. \tag{4}
$$

For further considerations, a special form of the density matrix is especially interesting, which is referred to as X form,[41,42] given by

$$\rho = \begin{pmatrix} \rho_{11} & 0 & 0 & \rho_{14} \\ 0 & \rho_{22} & \rho_{23} & 0 \\ 0 & \rho_{32} & \rho_{33} & 0 \\ \rho_{41} & 0 & 0 & \rho_{44} \end{pmatrix}. \tag{5}$$

Physically, the X form corresponds to a situation where all coherences between the ground state $|1\rangle$ and the single excitation states $|2\rangle$ and $|3\rangle$, and between $|2\rangle$, $|3\rangle$ and the double excitation state $|4\rangle$ are zero. The X form of the density matrix can be easily created by an appropriate initial preparation of a two-atom system. This form can be preserved during the evolution, or it can be created during the evolution.

Since the dipole-dipole interaction Ω_{12} couples the two atoms, the standard basis (4) is usually not the most convenient basis to work with, when describing the evolution of such a system. In this case, it is more convenient to include the dipole-dipole interaction into the Hamiltonian and rediagonalize it. As a result we get a different set of basis states that are particularly useful to describe the two-atom system. These are collective states of the system, or the Dicke states, defined as[38–40,43]

$$\begin{aligned} |g\rangle &= |g_1\rangle \otimes |g_2\rangle, \\ |s\rangle &= \frac{1}{\sqrt{2}} \left(|e_1\rangle \otimes |g_2\rangle + |g_1\rangle \otimes |e_2\rangle \right), \\ |a\rangle &= \frac{1}{\sqrt{2}} \left(|e_1\rangle \otimes |g_2\rangle - |g_1\rangle \otimes |e_2\rangle \right), \\ |e\rangle &= |e_1\rangle \otimes |e_2\rangle. \end{aligned} \tag{6}$$

It is interesting to note that the collective basis contains two states, $|s\rangle$ and $|a\rangle$ that are linear symmetric and antisymmetric superpositions of the product states, respectively. The most important is that the two states, $|s\rangle$ and $|a\rangle$, are in the form of maximally entangled states, or Bell states. The remaining two states, $|e\rangle$ and $|g\rangle$, are separable states.

The density matrix that has X form in the standard basis, has also X form in collective states

$$\rho = \begin{pmatrix} \rho_{gg} & 0 & 0 & \rho_{ge} \\ 0 & \rho_{ss} & \rho_{sa} & 0 \\ 0 & \rho_{as} & \rho_{aa} & 0 \\ \rho_{eg} & 0 & 0 & \rho_{ee} \end{pmatrix}, \tag{7}$$

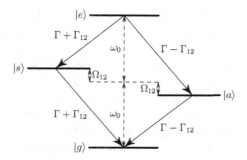

Figure 1.   Collective states and transition rates for the two-atom system

and this form is preserved during the evolution governed by the master Eq. (1). From the master Eq. (1), we get the following system of equations[40] for the matrix elements

$$\dot{\rho}_{ee} = -2\Gamma(1+N)\rho_{ee} + N\left[(\Gamma+\Gamma_{12})\rho_{ss} + (\Gamma-\Gamma_{12})\rho_{aa}\right]$$
$$\dot{\rho}_{ss} = (\Gamma+\Gamma_{12})\left[\rho_{ee} - (1+3N)\rho_{ss} - N\rho_{aa} + N\right]$$
$$\dot{\rho}_{aa} = (\Gamma-\Gamma_{12})\left[\rho_{ee} - N\rho_{ss} - (1+3N)\rho_{aa} + N\right] \qquad (8)$$
$$\dot{\rho}_{as} = -\left[\Gamma(1+2N) + 2i\Omega_{12}\right]\rho_{as}$$
$$\dot{\rho}_{ge} = -\Gamma(1+2N)\rho_{ge}$$

Solving Eqs. (8), we find all the matrix elements required for calculating the time evolution of the system, in particular, the entanglement evolution.

3. Entanglement measure

To quantify the entanglement, we need some measure of entanglement. In case of two qubits a popular and easy to calculate measure of entanglement is concurrence $\mathcal{C}$ introduced by Wootters.[29] The value of concurrence $\mathcal{C} = 0$ means that there is no entanglement, and the value $\mathcal{C} = 1$ means a maximally entangled state. For the X form of the density matrix of the system, the concurrence can be calculated analytically, and, in collective states, it is given by[7]

$$\mathcal{C}(t) = \max\left\{0, \mathsf{C}_1(t), \mathsf{C}_2(t)\right\}$$
$$\mathsf{C}_1(t) = 2|\rho_{ge}(t)| - \sqrt{\left[\rho_{ss}(t) + \rho_{aa}(t)\right]^2 - \left[2\mathrm{Re}\rho_{sa}(t)\right]^2} \qquad (9)$$
$$\mathsf{C}_2(t) = \sqrt{\left[\rho_{ss}(t) - \rho_{aa}(t)\right]^2 + \left[2\mathrm{Im}\rho_{sa}(t)\right]^2} - 2\sqrt{\rho_{ee}(t)\rho_{gg}(t)}$$

Inserting into Eqs. (9) the solutions to Eqs. (8), we find the values of $C_1(t)$ or $C_2(t)$, the criteria for entanglement. Whenever one of the two quantities becomes positive, there is a some degree of entanglement in the system. From Eqs. (9) it is clear that there is a competition between the inner 2×2 block of states $\{|s\rangle, |a\rangle\}$ and the outer 2×2 block of states $\{|g\rangle, |e\rangle\}$ in contributing to $C_1(t)$ and $C_2(t)$. There are two exclusive ways that lead to entanglement in the system. The competition introduces a sort of threshold conditions for obtaining entanglement, which lead to sometimes unexpected behavior of entanglement. One immediate and nice feature emerging from Eqs. (9), is the fact that, if the two-atom system is prepared in either symmetric or antisymmetric state, which both are maximally entangled states, the concurrence evolves in time as the population $\rho_{ss}(t)$ of the symmetric state or as the population $\rho_{aa}(t)$ of the antisymmetric state, respectively. These are exceptionally simple and striking results. In other situations the evolution of concurrence is much more complex, and it is discussed in the following sections.

4. Entanglement evolution: zero temperature reservoir

To discuss the entanglement evolution, we begin with a simple case of the reservoir being the vacuum of the electromagnetic modes, so that we put $N = 0$ in Eqs. (8). One can see from Eqs. (8) that the transition rates to and from the symmetric and antisymmetric states are modified by the collective damping Γ_{12}. Provided that $\Gamma_{12} > 0$, the transitions to and from the symmetric state occur with an enhanced rate $\Gamma + \Gamma_{12}$, whereas the transitions to and from the antisymmetric state occur with a reduced rate $\Gamma - \Gamma_{12}$. The collective damping Γ_{12}, given by Eq. (2), depends on the atomic separation. For small separations between the atoms, $\Gamma_{12} \approx \Gamma$, and then the state $|s\rangle$ becomes *superradiant* with a decay rate double that of the single atom Γ, whereas the state $|a\rangle$ becomes *subradiant*, with a decay rate of order $(kr_{12})\Gamma$ which vanishes in the limit of small distances $kr_{12} \ll 1$.

The set of coupled equations for the populations of the collective states can be easily solved, and the solution, valid for arbitrary initial conditions and $N = 0$, is given by[38,40]

$$\rho_{ee}(t) = \rho_{ee}(0)\, \mathrm{e}^{-2\Gamma t},$$

$$\rho_{ss}(t) = \rho_{ss}(0)\, \mathrm{e}^{-(\Gamma+\Gamma_{12})t} + \rho_{ee}(0)\frac{\Gamma + \Gamma_{12}}{\Gamma - \Gamma_{12}}\left[\mathrm{e}^{-(\Gamma+\Gamma_{12})t} - \mathrm{e}^{-2\Gamma t}\right], \quad (10)$$

$$\rho_{aa}(t) = \rho_{aa}(0)\, \mathrm{e}^{-(\Gamma-\Gamma_{12})t} + \rho_{ee}(0)\frac{\Gamma - \Gamma_{12}}{\Gamma + \Gamma_{12}}\left[\mathrm{e}^{-(\Gamma-\Gamma_{12})t} - \mathrm{e}^{-2\Gamma t}\right],$$

and $\rho_{gg}(t) = 1 - \rho_{ee}(t) - \rho_{ss}(t) - \rho_{aa}(t)$.

We see from Eqs. (10) that the decay of the populations depends strongly on the initial state of the system. When the system is initially prepared in the state $|s\rangle$, the population of the initial state decays exponentially with an enhanced rate $\Gamma + \Gamma_{12}$, while the initial population of the antisymmetric state $|a\rangle$ decays with a reduced rate $\Gamma - \Gamma_{12}$. This occurs because the photons emitted from the excited atom can be absorbed by the atom in the ground state, so that the photons do not escape immediately from the system. For a general initial state that includes the state $|e\rangle$, the populations of the symmetric and the antisymmetric states do not decay with a single exponential.

Similarly, equations of motion for the coherences in Eqs. (8), are straightforward to solve, giving

$$\rho_{sa}(t) = \rho_{sa}(0) \, e^{-(\Gamma + 2i\Omega_{12})t},$$
$$\rho_{ge}(t) = \rho_{ge}(0) \, e^{-(\Gamma - 2i\omega_0)t}. \tag{11}$$

Having at hand solutions (10) and (11), it is just a matter of inserting them into expressions (9) to get analytical formulas describing the criteria for entanglement.

4.1. *Creation of entanglement*

Let us start our discussion of entanglement evolution by assuming that initially one atom is excited and the other one is in its ground state, to be more specific, let $|\Psi(0)\rangle = |e_1\rangle \otimes |g_2\rangle$. Of course, this initial state is separable, so there is no entanglement initially. In terms of collective states, it means that the only non-zero initial matrix elements are: $\rho_{ss} = \rho_{aa} = \rho_{sa} = \rho_{as} = 1/2$. It is easy to verify, that for such initial state the concurrence is equal to[44]

$$C(t) \equiv C_2(t) = e^{-\Gamma t} \sqrt{\sinh^2(\Gamma_{12}t) + \sin^2(2\Omega_{12}t)}. \tag{12}$$

Formula (12) has two parts, an oscillatory part that oscillates with frequency $2\Omega_{12}$, and non-oscillatory part that depends on Γ_{12}. It is evident that for $t > 0$ the concurrence (12) becomes positive, *i.e.*, entanglement is created by spontaneous emission from the system. If $\Omega_{12} \gg \Gamma$ concurrence exhibits oscillations. This is illustrated in Fig. 2, where concurrence is plotted according to (12) for $r_{12} = \lambda/12$. Envelopes of this oscillatory function are given by $\rho_{aa}(t) + \rho_{ss}(t)$ and $\rho_{aa}(t) - \rho_{ss}(t)$, which are plotted for reference. Since $\rho_{ss}(t)$ decays much faster than $\rho_{aa}(t)$, eventually, only the antisymmetric state survives, and for long times, the concurrence is

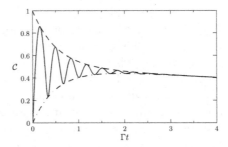

Figure 2.   Concurrence $\mathcal{C}(t)$ for initially one atom excited, according to Eq. (12), (solid). Envelopes are: $\rho_{aa}(t)+\rho_{ss}(t)$ (dashed), $\rho_{aa}(t)-\rho_{ss}(t)$ (dashed-dotted). Parameters are: $r_{12} = \lambda/12$ and $N = 0$.

equal to the population $\rho_{aa}(t)$. The antisymmetric state plays a crucial role in creating entanglement via spontaneous emission. If the antisymmetric state is eliminated from the evolution, as it is the case in the Dicke model, or small sample model, entanglement cannot be created by spontaneous emission.

4.2. *Sudden death of entanglement*

One of the characteristic features of entanglement in a bipartite system is an abrupt disappearance of initial entanglement at finite time, the effect for which Yu and Eberly[12,21,22] introduced the name *sudden death of entanglement*. They considered a two qubit system prepared initially in a state which is diagonal in the collective states basis, and the initial density matrix has the form

$$\rho(0) = \frac{1}{3}\begin{pmatrix} 1-\alpha & 0 & 0 & 0 \\ 0 & 2 & 0 & 0 \\ 0 & 0 & 0 & 0 \\ 0 & 0 & 0 & \alpha \end{pmatrix}, \tag{13}$$

where $0 \leq \alpha \leq 1$. With this initial state the concurrence is described by a simple formula

$$\mathcal{C}(t) = \mathsf{C}_2(t) = |\rho_{ss}(t) - \rho_{aa}(t)| - 2\sqrt{\rho_{gg}(t)\rho_{ee}(t)}, \tag{14}$$

where the solutions (10) are to be inserted. The results are illustrated in Fig. 3. Figure (a) presents the evolution of concurrence for the interatomic distance $r_{12} = 10\lambda$, which reproduces exactly the famous picture of Yu

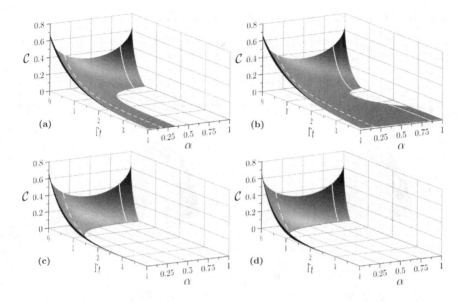

Figure 3.   Concurrence evolution for the vacuum reservoir ($N = 0$) and the initial state (13): (a) $r_{12} = 10\lambda$, (b) $r_{12} = \lambda/2$, (c) $r_{12} = \lambda/4$, (d) $r_{12} = \lambda/12$.

and Eberly[12,21,22] for atoms coupled to independent reservoirs. The interatomic distance $r_{12} = 10\lambda$ appears to be large enough to consider the common reservoir as two independent reservoirs, so both atoms evolve independently. To emphasize the two different types of evolution, *i.e.*, sudden death of entanglement and asymptotic decay, we mark in the figure two trajectories representing the two evolutions: the solid line which represents the sudden death of entanglement and the dashed line which shows the asymptotic decay. It is seen that for $\alpha < 1/3$ evolution is asymptotic and for $\alpha > 1/3$ sudden death of entanglement is observed.

The situation changes radically when the interatomic distance becomes shorter and atoms behave collectively. It is shown in Fig. 3 (c)–(d) that for distances shorter than $\lambda/4$ sudden death appears practically for the whole range of α. Only for $\alpha = 0$ we have asymptotic behavior because in this case the concurrence follows the exponential decay of $\rho_{ss}(t)$. In Fig. 3 (b) an interesting effect of sudden death followed by a revival of entanglement is seen. So, the collective behavior of the atoms makes the entanglement evolution dependent on the interatomic distance and it becomes quite different from that for independent atoms.

4.3. *Sudden death and revival of entanglement*

Another interesting example of entanglement evolution takes place when the two atoms are prepared initially in the state

$$|\Psi(0)\rangle = \sqrt{p}|e\rangle + \sqrt{1-p}|g\rangle \qquad (15)$$

This state is the superposition of the states: one with both atoms excited

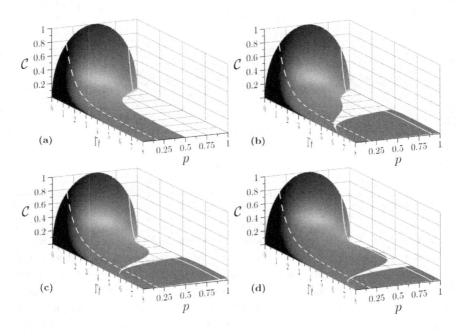

Figure 4.   Concurrence evolution for the vacuum reservoir ($N = 0$) and the initial state (15): (a) $r_{12} = 10\lambda$, (b) $r_{12} = \lambda/4$, (c) $r_{12} = \lambda/8$, (d) $r_{12} = \lambda/20$.

and the other with both atoms in their ground states. Again, for independent atoms clear splitting into two regions is seen, we find the sudden death of entanglement for $p > 0.5$, and the asymptotic evolution for $p < 0.5$, as shown in Fig. 4 (a). However, for interatomic distances shorter than the wavelength of emitted light, the structure of entanglement evolution exhibits very interesting features, as shown in Fig. 4 (b)–(d), which include a broad range of entanglement sudden death accompanied by a subsequent revival. Especially interesting is the evolution of entanglement for values of $p \sim 0.9$, which is shown in Fig. 4 (d), and in more details in Fig. 5. As it is clear, the entanglement experiences sudden death and revival not just once, but twice.[15] This happens for the interatomic distance $r_{12} = \lambda/20$,

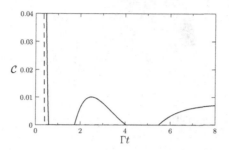

Figure 5. Concurrence evolution for the initial state (15) and interatomic distance $r_{12} = \lambda/20$; dashed line shows the evolution for independent atoms.

which is quite short, but it illustrates how important for the entanglement evolution can be the collective behavior of the atoms. For reference, the dashed curve in Fig. 5 shows the evolution for independent atoms.

At this point it can be interesting to compare the results obtained above, which were obtained for the model taking into account the collective damping (2) and the dipole-dipole interaction (3), both depending on the interatomic separation, to the Dicke model, or the small sample model, which is based on the assumption that the two atoms are very close to each other, so that $\Gamma_{12} = \Gamma$, and the dipole-dipole interaction can be ignored ($\Omega_{12} = 0$). In this case the antisymmetric state (singlet state) does not evolve in time. The evolution is restricted to the three triplet states ($\{|g\rangle, |s\rangle, |e\rangle\}$), and it is governed by the equations

$$\rho_{ee}(t) = \rho_{ee}(0)e^{-2\Gamma t}$$
$$\rho_{ss}(t) = \rho_{ss}(0)e^{-2\Gamma t} + 2\Gamma t\,\rho_{ee}(0)e^{-2\Gamma t} \qquad (16)$$
$$\rho_{ge}(t) = \rho_{ge}(0)e^{-(\Gamma-2i\omega_0)t}.$$

It is worth to notice the non-exponential evolution of $\rho_{ss}(t)$, if $\rho_{ee}(0) \neq 0$. The concurrence evolution for the Dicke model and the initial state (15) is plotted in Fig. 6. The difference between the Dicke model and the extended model shown in Fig. 4 is clearly visible. Entanglement evolution for the Dicke model and a common structured reservoir has been discussed in Ref. 34. They found revivals of entanglement for non-Markovian reservoir with Lorentzian spectrum. In the limit of very broad spectrum of the reservoir the result should reproduce that for Markovian reservoir, our Fig. 6

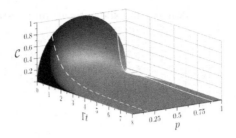

Figure 6.   Concurrence evolution for the vacuum reservoir ($N = 0$) and the initial state (15) for the Dicke model

really agrees with the corresponding figure in Ref. 34. We want to emphasize here that the Dicke model, described by Eqs. (16) and illustrated in Fig. 6, essentially differs from the extended model used in this paper.

4.4. *Sudden birth of entanglement*

As we have seen already, the collective spontaneous emission can create entanglement in the two-atom system in a smooth way, when the atoms initially start from a specific product state. If there is some entanglement in the system initially, it can disappear abruptly in a finite time, or it can decay asymptotically. Here, we want to discuss another interesting feature of entanglement evolution. Let us consider another initial product state of

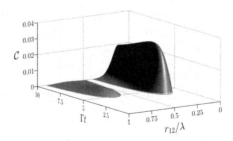

Figure 7.   Concurrence as a function of Γt and r_{12}, for vacuum reservoir ($N = 0$) and initially both atoms excited

the two-atom system, such that initially both atoms are excited, so the only non-zero element of the atomic density matrix is $\rho_{ee}(0) = 1$. In this case, according to solutions (10), we find that the density matrix has a diagonal form, and when we check the criteria for entanglement (9), we easily discover that the only contribution to entanglement, if any, can come from $C_2(t)$, which is given by Eq. (14), together with the solutions (10) and appropriate initial conditions. It is clear from Eq. (14) that entanglement can result solely from unequal populations of the symmetric and antisymmetric states. When the system starts from the state $|e\rangle$, there are two channels of spontaneous emission: the fast one through the state $|s\rangle$, and the slow one through the state $|a\rangle$, as indicated in Fig. 1. Due to the difference in the emission rates, the difference between the populations builds up, but it is counteracted by the accumulation of the population $\rho_{gg}(t)$ in the ground state $|g\rangle$. In effect, we get another "sudden" phenomenon in the entanglement evolution, which we refer to as *sudden birth of entanglement*.[17] To visualize the behavior of the concurrence for this case, we plot in Fig. 7 the concurrence C as a function of the dimensionless evolution time Γt and the interatomic distance r_{12}. We see that there is no entanglement at early times of evolution, but suddenly after a finite time an entanglement emerges. However, as it is seen, it happens only for sufficiently small interatomic distances r_{12}, for which atoms behave collectively. Once again, the collective behavior of the atoms appears to be crucial for observing this phenomenon.

Among the states having the X form of the density matrix, which are of great importance in quantum information theory, are Werner states, with the density matrix of the form

$$\rho(0) = p|s\rangle\langle s| + (1-p)\frac{\mathbb{I}}{4}, \tag{17}$$

where $\mathbb{I}$ is the 4×4 identity matrix. Werner states represent a mixture of the maximally entangled state and the isotropic state. In our case the maximally entangled state is the symmetric state $|s\rangle$, which contributes to the mixture with the probability p. Werner states, depending on the value of p, in a sense, interpolate between entangled states and separable states. For $p = 1$, we have a maximally entangled state with concurrence $C = 1$, while for $p < 1/3$ the state is separable. For $p > 1/3$ the concurrence has the value $C(0) = (3p-1)/2$. Such states are excellent examples for illustrating all the features of entanglement evolution for the system of two atoms coupled to a common reservoir. It is convincingly shown in Fig. 8, where the concurrence evolution is plotted as a function of Γt and the probability p, for various values of the interatomic distance r_{12}. For independent atoms, Fig. 8 (a), as before, two trajectories are marked, one showing the sudden death of entanglement (solid line) and the other showing the asymptotic evolution (dashed line). The same trajectories are marked in the remaining figures,

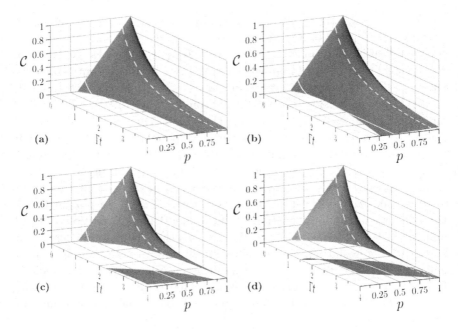

Figure 8.   Concurrence evolution for vacuum reservoir ($N = 0$) and initial state (17): (a) $r_{12} = 10\lambda$, (b) $r_{12} = \lambda/2$, (c) $r_{12} = \lambda/4$, (d) $r_{12} = \lambda/12$.

but, as it is seen, they no longer represent sudden death and asymptotic evolutions. As the atoms behave collectively, the concurrence exhibits much more interesting behavior. In Fig. 8 (b) we see that the trajectory that experienced sudden death, after some time exhibits revival of entanglement. The revival is also seen in figures (c) and (d). Moreover, for independent atoms there is no entanglement for $p < 1/3$, and from figures (c) and (d) we see the presence of entanglement at later times, which means the sudden birth of entanglement. However, for the trajectory representing initially asymptotic evolution, sudden death and subsequent revival of entanglement appears. Generally, we can say that the sudden death is a rather common phenomenon when the two atoms behave collectively, but the sudden death can be followed by a revival, after which asymptotic evolution can take place.

5. Entanglement evolution: thermal reservoir

Now, we want to discuss the situation when the reservoir is not the vacuum, or zero-temperature reservoir, but it is a thermal reservoir with the mean number of photons N different from zero. To find the evolution of concurrence it is now necessary to solve the full version, with $N \neq 0$, of Eqs. (8). Although analytical solutions are possible, they are rather complicated and

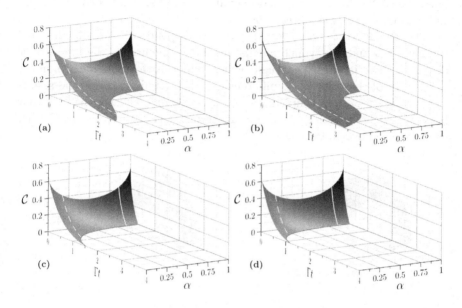

Figure 9.   Same as Fig. 3 but for thermal reservoir with $N = 0.01$.

we do not give them here. In further calculations of the concurrence we will rely on numerical solutions.

However, before we calculate the time evolution of concurrence, we find the steady state solutions to Eqs. (8), which are given by

$$\rho_{ee}(\infty) = \frac{N^2}{(2N+1)^2}$$

$$\rho_{ss}(\infty) = \rho_{aa}(\infty) = \frac{N(N+1)}{(2N+1)^2} \tag{18}$$

$$\rho_{gg}(\infty) = \frac{(N+1)^2}{(2N+1)^2},$$

and the steady state values of coherences $\rho_{as}(\infty)$ and $\rho_{eg}(\infty)$ are zero. From the solutions (18), it is immediately seen that both $C_1(t)$ and $C_2(t)$, given by Eqs. (9), take negative values as $t \to \infty$. This means that there is no entanglement in the system for $t \to \infty$. So, no matter how large the degree of entanglement could be in the system at earlier times, there must be a finite time t_d at which entanglement disappears. Of course, the death time t_d depends on the value of the mean number of photons N of the reservoir, but for any $N > 0$, no matter how small it could be, there is entanglement sudden death for thermal reservoir.

The steady state solutions (18) do not depend on the collective parameters Γ_{12} and Ω_{12}, which means that independently of the interatomic distance, there is always entanglement sudden death if the mean number of photons of the reservoir is different from zero. This confirms the results found earlier for atoms coupled to separate reservoirs.[36,37] For the long-time behavior of the system in a thermal reservoir, it is not important whether the atoms behave collectively or not.

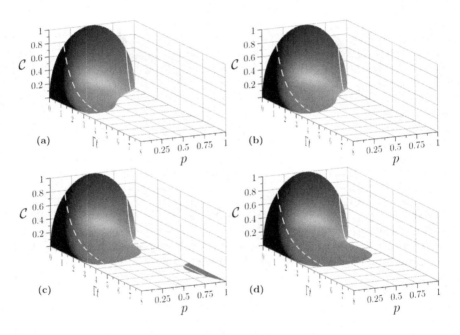

Figure 10.   Same as Fig. 4 but for thermal reservoir with $N = 0.01$.

To illustrate entanglement evolution in a thermal reservoir, we calculate the concurrence for the initial states that we used earlier to discuss the evolution of the two-atom system in the vacuum. In Fig. 9 we plot the concurrence for initial state (13), discussed by Yu and Eberly[12,21,22] for atoms embedded in a common thermal reservoir with the mean number of

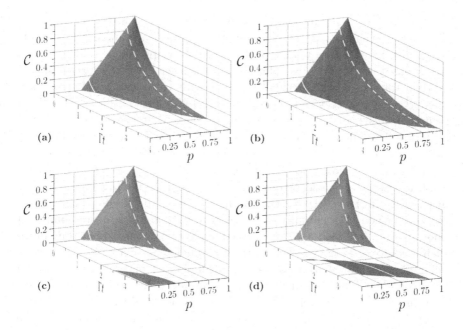

Figure 11.   Same as Fig. 8 but for thermal reservoir with $N = 0.01$.

photons $N = 0.01$. The remaining parameters are the same as in Fig. 3. It is evident from Fig. 9 that there is sudden death of entanglement for the whole range of α values. The death time t_d depends on the interatomic distance and the value of α in a rather complicated way, but sudden death of entanglement is clearly evident, and there is no asymptotic evolution. For small interatomic distances, as a rule, the death time becomes shorter for shorter distances.

In Fig. 10 the concurrence evolution is shown for the initial state (15), for the same values of interatomic distances as in Fig. 4, but for the thermal reservoir with the mean number of photons $N = 0.01$. We see that the evolution has drastically changed with respect to the vacuum reservoir. Again, as in the previous case, the sudden death of entanglement is clearly visible. Moreover, the revivals of entanglement that are so evident in Fig. 4, disappeared almost completely, the remnants are still visible in figure (c), for $r_{12} = \lambda/8$, but they will also disappear if the number of photons will increase. It is also seen that, for the initial state (15), in contrast to the initial state (13) illustrated in Fig. 9, the shorter interatomic distance does not mean the shorter death time of entanglement. The dependence on the interatomic distance is in this case reversed. However, in any case, the death time becomes shorter as the mean number of photons N increases, but this is not illustrated in the figures presented in the paper.

Similarly, in Fig. 11, we plot the concurrence behavior for the initial Werner state (17) and thermal reservoir with the mean number of photons $N = 0.01$. Direct comparison of Fig. 8 and Fig. 11 shows that there is sudden death of entanglement for the cases that previously exhibited asymptotic evolution. In cases, for which sudden death appeared due to the collective behavior of the atoms, in thermal reservoir the death time is shortened. Moreover, for the interatomic distances for which collective evolution of the atoms leads to the revival or sudden birth of entanglement in the vacuum reservoir, for the thermal reservoir both effects are gradually diminished.[45] Eventually, for a reservoir with a higher mean number of photons, they will be completely erased, and the only effect that remains in such a reservoir is the sudden death of entanglement, with the death time being shorter and shorter as the mean number of photons increases. All the examples of entanglement evolution discussed here convincingly show that the collective behavior of the atoms, discussed in the paper, leads to a quite interesting evolution of entanglement, with new features, when the two atoms are embedded in a common reservoir being in the vacuum state. When the reservoir has non-zero temperature, the structure of entanglement evolution simplifies, and eventually becomes quite simple, showing always the sudden death of entanglement. No asymptotic evolution is possible when the temperature is non-zero, no matter whether the atoms are coupled to separate reservoirs or to a common reservoir. It shows that the zero-temperature reservoir plays a special role in entanglement evolution.

6. Conclusion

We have discussed the dynamics of entanglement in a two-atom system interacting with a common reservoir at zero temperature (vacuum) as well as at finite temperature (thermal reservoir). The evolution of the system is described by the Lehmberg-Agarwal Markovian master equation, which takes into account collective behavior of the atoms. The collective spontaneous emission is a source of entanglement for initially separable states. The entanglement can be created continuously from the beginning of the evolution, or it can be created abruptly after a finite time of the evolution that took place without entanglement. The latter is referred to as entanglement sudden birth. The entanglement which is already present in the system can disappear in a finite time, which is the effect of entanglement sudden death, but it can also revive after some time. For some initial states and vacuum reservoir, entanglement can decay asymptotically (for $t \to \infty$). Generally, for atoms behaving collectively, the structure of the entanglement evolution appears to be quite rich. However, this rich structure of the evolution is degraded for the non-zero temperature reservoir, for which the entanglement sudden death becomes the standard feature of the evolution, and only in the limit $N \to 0$ the asymptotic decay of entanglement is possible. This means that, in real physical situations of finite temperature reservoirs, there is always entanglement sudden death. On the other hand, entanglement sudden birth created by correlated atoms appears only for reservoirs at sufficiently low temperatures, and it disappears at higher temperatures. We have made a comparison of the concurrence evolution for a number of initial states of the two-atom system to show the variety of effects that can be observed in the evolution.

Acknowledgements

I acknowledge continuous and effective collaboration with Zbigniew Ficek on the subject of this work.

Bibliography

1. R. Horodecki, P. Horodecki, M. Horodecki and K. Horodecki, *Rev. Mod. Phys.* **81**, p. 865 (2009).
2. K. Życzkowski, P. Horodecki, M. Horodecki and R. Horodecki, *Phys. Rev. A* **65**, p. 012101 (2001).
3. L. Jakóbczyk, *J. Phys. A* **35**, p. 6383 (2002).
4. D. Braun, *Phys. Rev. Lett.* **89**, p. 277901 (2002).
5. M. S. Kim, J. Lee, D. Ahn and P. L. Knight, *Phys. Rev. A* **65**, p. 040101 (2002).
6. A. M. Basharov, *JETP Lett* **75**, p. 123 (2002).
7. Z. Ficek and R. Tanaś, *J. Mod. Opt.* **50**, p. 2765 (2003).
8. X. X. Yi, C. S. Yu, L. Zhou and H. S. Song, *Phys. Rev. A* **68**, p. 052304 (2003).
9. L. Jakóbczyk and A. Jamróz, *Phys. Lett. A* **318**, p. 318 (2003).
10. S. Nicolosi, A. Napoli, A. Messina and F. Petruccione, *Phys. Rev. A* **70**, p. 022511 (2004).
11. L. Jakóbczyk and A. Jamróz, *Phys. Lett. A* **333**, p. 35 (2004).
12. T. Yu and J. H. Eberly, *Phys. Rev. Lett.* **93**, p. 140404 (2004).
13. P. J. Dodd, *Phys. Rev. A* **69**, p. 052106 (2004).
14. L. Jakóbczyk and A. Jamróz, *Phys. Lett. A* **347**, p. 180 (2005).
15. Z. Ficek and R. Tanaś, *Phys. Rev. A* **74**, p. 024304 (2006).
16. A. Jamróz, *J. Phys. A* **39**, p. 7727 (2006).

17. Z. Ficek and R. Tanaś, *Phys. Rev. A* **77**, p. 054301 (2008).
18. F. Verstraete, M. M. Wolf and I. Cirac, *Nature Physics* **5**, p. 633 (2009).
19. K. Roszak, P. Horodecki and R. Horodecki, *Phys. Rev. A* **81**, p. 042308 (2010).
20. Z. Ficek, *Front. Phys. China* **5**, p. 26 (2010).
21. J. H. Eberly and T. Yu, *Science* **316**, p. 555 (2007).
22. T. Yu and J. H. Eberly, *Science* **323**, p. 598 (2009).
23. M. P. Almeida, F. de Melo, M. Hor-Meyll, A. Salles, S. P. Walborn, P. H. Souto Ribeiro and L. Davidovich, *Science* **316**, p. 579 (2007).
24. B. Bellomo, R. Lo Franco and G. Compagno, *Phys. Rev. Lett.* **99**, p. 160502 (2007).
25. X. Cao and H. Zheng, *Phys. Rev. A* **77**, p. 022320 (2008).
26. F. Wang, Z. Zhang and R. Liang, *Phys. Rev. A* **78**, p. 062318 (2008).
27. R. Tanaś and Z. Ficek, *J. Opt. B* **6**, p. S610 (2004).
28. D. Mundarain and M. Orszag, *Phys. Rev. A* **75**, p. 040303 (2007).
29. W. K. Wootters, *Phys. Rev. Lett* **80**, p. 2245 (1998).
30. A. Peres, *Phys. Rev. Lett.* **77**, p. 1413 (1996).
31. M. Horodecki, P. Horodecki and R. Horodecki, *Phys. Lett. A* **223**, p. 1 (1996).
32. G. H. Yang and L. Zhu, *Eur. Phys. J. D* **47**, p. 277 (2008).
33. C. E. López, G. Romero, E. Lastra, E. Solano and J. C. Retamal, *Phys. Rev. Lett.* **101**, p. 080503 (2008).
34. L. Mazzola, S. Maniscalco, J. Piilo, K.-A. Suominen and G. Garraway, *Phys. Rev. A* **79**, p. 042302 (2009).
35. K. Härkönen, F. Plastina and S. Maniscalco, *Phys. Rev. A* **80**, p. 033841 (2009).
36. M. Ikram, F. li Li and M. S. Zubairy, *Phys. Rev. A* **75**, p. 062336 (2007).
37. A. Al-Qasimi and D. F. V. James, *Phys. Rev. A* **77**, p. 012117 (2008).
38. R. H. Lehmberg, *Phys. Rev. A* **2**, p. 883 (1970).
39. G. S. Agarwal, *Quantum Statistical Theories of Spontaneous Emission and their Relation to other Approaches*, Springer Tracts in Modern Physics, Vol. 70 (Springer, Berlin, 1974).
40. Z. Ficek and R. Tanaś, *Phys. Reports* **372**, p. 369 (2002).
41. T. Yu and J. H. Eberly, *Quant. Inf. and Comp.* **7**, p. 459 (2007).
42. B. Corn and T. Yu, *Quant. Inf. Process.* **8**, p. 565 (2009).
43. R. H. Dicke, *Phys. Rev.* **93**, p. 99 (1954).
44. R. Tanaś and Z. Ficek, *J. Opt. B* **6**, p. S90 (2004).
45. R. Tanaś and Z. Ficek, *Physica Scripta* (2010), in print.

Open system dynamics of simple collision models

M. Ziman*

V. Bužek

*Institute of Physics, Slovak Academy of Sciences,
Dúbravská cesta 9, 845 11 Bratislava, Slovakia
and
Faculty of Informatics, Masaryk University,
Botanická 68a, 602 00 Brno, Czech Republic
* E-mail: ziman@savba.sk*

A simple collision model is employed to introduce elementary concepts of open
system dynamics of quantum systems. In particular, within the framework of
collision models we introduce the quantum analogue of thermalization process
called quantum homogenization and simulate quantum decoherence processes.
These dynamics are driven by partial swaps and controlled unitary collisions,
respectively. We show that collision models can be used to prepare multipartite
entangled states. Partial swap dynamics generates W-type of entanglement
saturating the CKW inequalities, whereas the decoherence collision models
creates GHZ-type of entangled states. The considered evolution of a system
in a sequence of collisions is described by a discrete semigroup $\mathcal{E}_1, \ldots, \mathcal{E}_n$.
Interpolating this discrete points within the set of quantum channels we derive
for both processes the corresponding Linblad master equations. In particular,
we argue that collision models can be used as simulators of arbitrary Markovian
dynamics, however, the inverse is not true.

Keywords: quantum dynamics; quantum master equations; quantum thermalization; quantum decoherence; entanglement dynamics.

1. Open system dynamics

The goal of these lectures is to present the elementary ideas and features
of dynamics of open quantum systems by analyzing the properties of the
simplest collision model we can think of. The material is based on papers[1-4]
on collision models coauthored by authors. It is not considered as a review
paper on collision models or open system dynamics.

It turns out it is surprisingly useful in physics to distinguish the concepts
of isolated and open systems. The concept of isolated systems represents a

simplification of physical reality, which serves as a playground for a clear formulation of elementary physical principles. It is postulated that the dynamics of isolated quantum systems is driven by *Schrödinger equation*[5]

$$i\hbar\frac{d}{dt}\psi_t = H_t\psi_t\,, \tag{1}$$

where H_t is a hermitian operator called *Hamiltonian*. As a consequence it follows that the state transformations are unitary, i.e. $\psi_t \to \psi_{t'} = U_{t\to t'}\psi_t$, where $U_{t\to t'}U_{t\to t'}^\dagger = U_{t\to t'}^\dagger U_{t\to t'} = I$. If H_t is time-independent, i.e. $H_t \equiv H$, then $U_{t\to t'} = \exp\left[-\frac{i}{\hbar}H(t'-t)\right]$. We can define $U_t = \exp\left(-\frac{i}{\hbar}Ht\right)$ and write $U_{t\to t'} = U_{t'}U_t^\dagger$, $\psi_t = U_t\psi_0$.

Interactions between a system under consideration and its environment are responsible for the violation of system's (dynamical) isolation. The joint evolution of the system and the environment will be still unitary, however, the system itself undergoes a different dynamics. The point is that these interactions, because of the complexity of the environment, are typically out of our control. And our wish is to invent models of system-environemt interactions capturing faithfully the key properties of the dynamics of the system alone. For a general discussion on the models of open system dynamics we refer to.[6–8] In what follows we will restrict our analysis to a particular toy model of the open system dynamics.

In the rest of this section we will introduce the model and elementary properties of quantum channels. In the following two sections we will discuss thermalization and decoherence processes within this model. In Section IV we will focus on entanglement created in the discussed collision models and finally, in the Section V we will derive master equations for these collision processes.

1.1. *Simple collision model*

Let $\mathcal{H}$ be the Hilbert space of the system of interest. States are identified with elements of the set of density operators (positive operators of unit trace)

$$\mathcal{S}(\mathcal{H}) = \{\varrho : \varrho \geq O,\ \mathrm{tr}\varrho = 1\}\,. \tag{2}$$

We will assume that environment consists of huge number of particles, each of them described by the same Hilbert space $\mathcal{H}$. In a sense, it forms a reservoir of particles. Moreover, we assume that initially they are all in the same state ξ, i.e. the initial state of the environment/reservoir is $\omega = \xi^{\otimes N}$. In each time step the system interacts with a single environment's particle.

The interaction is described by a unitary operator U acting on joint Hilbert space $\mathcal{H} \otimes \mathcal{H}$. Thus, the one-step system's evolution is described by a map

$$\varrho \to \varrho' = \mathcal{E}[\varrho] = \mathrm{tr}_{\mathrm{env}}[U(\varrho \otimes \xi)U^\dagger], \tag{3}$$

where ϱ is the initial state of the system and $\mathrm{tr}_{\mathrm{env}}$ denotes the partial trace over the environment. In each time step a system undergoes a collision with a single particle from the reservoir. In a realistic scenarios these collisions are happening randomly, but we will assume that each reservoir's particle interacts with the system at most once (see Fig. 1). Thus, the system evolution is driven by a sequence of independent collisions. After the nth collision we get

$$\varrho_0 \to \varrho_n' = \mathrm{tr}_{\mathrm{env}}[U(\varrho_{n-1} \otimes \xi)U^\dagger] = \mathcal{E}[\varrho_{n-1}] = \mathcal{E}^n[\varrho_0]. \tag{4}$$

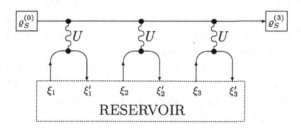

Figure 1.   A simple collision model.

1.2. *Quantum channels*

The mapping $\mathcal{E}$ defined on the set of all operators $\mathcal{L}(\mathcal{H})$ satisfies the following properties:

(1) $\mathcal{E}$ is linear, i.e. $\mathcal{E}[X + cY] = \mathcal{E}[X] + c\mathcal{E}[Y]$ for all $X, Y \in \mathcal{L}(\mathcal{H})$ and all complex numbers c.

(2) $\mathcal{E}$ is trace-preserving, i.e. $\mathrm{tr}\mathcal{E}[X] = \mathrm{tr}X$ for all operators X with finite trace.

(3) $\mathcal{E}$ is completely positive, i.e. $(\mathcal{E} \otimes \mathcal{I})[\Omega] \geq O$ for all positive operators $\Omega \in \mathcal{L}(\mathcal{H} \otimes \mathcal{H})$.

Due to *Stinespring dilation theorem* the inverse is also true. If a mapping $\mathcal{E}$ satisfies above three conditions, there is a unitary operator U and a state

ξ such that the Eq. (3) holds. Such completely positive trace-preserving linear maps we call *channels*.

For any state $\xi \in \mathcal{S}(\mathcal{H})$ there exist a unit vector $\Psi \in \mathcal{H} \otimes \mathcal{H}$ such that $\xi = \mathrm{tr}_2 |\Psi\rangle\langle\Psi|$, where tr_2 is the partial trace over the second subsystem. We call Ψ a *purification* of ξ. It is straightforward to verify that the following identity holds

$$\mathcal{E}[\varrho] = \mathrm{tr}_{\mathrm{env}}[U(\varrho \otimes \xi)U^\dagger] = \mathrm{tr}_{\mathrm{env}'}[(U \otimes I)(\varrho \otimes |\Psi\rangle\langle\Psi|)(U \otimes I)^\dagger]. \quad (5)$$

Let φ_j is an orthonormal basis on $\mathcal{H} \otimes \mathcal{H}$ and define linear operators $A_j = \langle\varphi_j|(U \otimes I)\Psi\rangle$ acting on the system. Then the right-hand side of the above equation reads

$$\mathcal{E}[\varrho] = \sum_j \langle\varphi_j|(U \otimes I)\Psi\rangle \varrho \langle(U \otimes I)\Psi|\varphi_j\rangle = \sum_j A_j \varrho A_j^\dagger. \quad (6)$$

By definition the so-called *Kraus operators* A_j satisfy the normalization $\sum_j A_j^\dagger A_j = I$. Any mapping of the Kraus form $\mathcal{E}[\varrho] = \sum_j A_j \varrho A_j^\dagger$ for arbitrary set of operators $\{A_1, A_2, \dots\}$ fulfilling the normalization constraint determines a valid quantum channel.

Channels with a single Kraus operator $\mathcal{E}[\varrho] = A\varrho A^\dagger$ are *unitary channels*, because the normalization $A^\dagger A = I$ implies that A is unitary. Let us note that Kraus representation and also the number of Kraus operators are not unique. For more details we refer to.[9,10]

2. Quantum homogenization as an analogue to thermalization

The 0th law of thermodynamics postulates that an object brought in a contact with a reservoir of temperature T will equalize its temperature with the reservoir's temperature. This process is known as the *process of thermalization*. Our attempt is to design a quantum analogue of this process.

The concept of temperature is the key ingredient of the thermalization process. In fact, as a consequence it is the temperature that describes a state of the system, that is, it contains the relevant part of the information needed for the investigation of the system's properties. What is the quantum analogue of the temperature? There are several ways how this concept can be introduced. One can use the connection between temperature and the entropy, or energy per single particle. By choosing such analogue, the thermalization can be defined as a process that leads a system with a given temperature T_S to the temperature of the reservoir T_ξ. In classical theory this process is achieved because of mutual interactions/collisions between the particles. In the process of collision they mutually (partly) exchange their microscopic properties (energy, momentum). Subsequently, the temperature of the system being in a contact with the reservoir is changing and finally (approximately) equalize T_ξ. Let us note that we have used the temperature as a property that can be attributed even to a single particle, however, such definition is meaningful only in the statistical sense.

In quantum theory the state of a quantum system is described with a density operator. This operator contains all the information. Therefore, we choose single-particle density operator to play the role of the quantum temperature in our quantum analogue. So the goal is to design a model of the dynamics of the system+reservoir such that the initial state of the system particle ϱ_S evolves into the state of the reservoir particles ξ. Because of the property that finally all particles are described by the same state ξ we will refer to this process as to *homogenization* instead of the thermalization. In fact this notion is more appropriate to characterize the features of our model.

Ideally the homogenization should implement the following transformation

$$\varrho_S \otimes \xi \otimes \cdots \otimes \xi \mapsto \xi_S \otimes \xi \otimes \cdots \otimes \xi, \tag{7}$$

for all ϱ and ξ. For finite reservoirs this process will perform $N \to N+1$ cloning transformation, which would violate the quantum no-cloning theorem.[11,12] Therefore, this process cannot be achieved in any dynamical model respecting the quantum theory. We will relax our requirements and investigate whether the homogenization can be implemented approximately via a sequence of (independent) collisions. Moreover, if initially $\varrho = \xi$, meaning there is nothing to homogenize, then the evolution is trivial. Let us summarize our assumptions and problem.

Definition 2.1. We call an interaction U a generator of δ-homogenization $\delta > 0$ is the following conditions are satisfied

- *Trivial homogenization.* The joint state of the system and the environment after nth interaction reads

$$\Omega_n = U_n \cdots U_1 (\varrho_S \otimes \xi^{\otimes n}) U_1^\dagger \cdots U_n^\dagger, \tag{8}$$

where $U_j = U_{S,j} \otimes I_{\text{env} \setminus \{j\}}$ describes the jth interaction ($I_{\text{env} \setminus \{j\}}$ is the identity operator on the environment excluding the jth particle). The for all $\xi \in \mathcal{S}(\mathcal{H})$

$$\xi_S \otimes \xi \otimes \cdots \otimes \xi \mapsto \Omega_n = \xi_S \otimes \xi \otimes \cdots \otimes \xi. \tag{9}$$

- *Convergence.* Let $\varrho_S^{(n)}$ be the state of the system after nth collision. Then there exist N_δ such that for all $n > N_\delta$

$$D(\varrho_S^{(n)}, \xi) \leq \delta. \tag{10}$$

- *Stability of the reservoir.* Let ξ_j' be the state of the jth reservoir's particle after the its interaction with the system. Then

$$D(\xi_j', \xi) \leq \delta, \tag{11}$$

for all j.

Our task is simple: find U satisfying the above properties.

2.1. *Trivial homogenization*

Let us denote by $\mathcal{H}_\pm$ the symmetric and antisymmetric subspaces of $\mathcal{H} \otimes \mathcal{H}$. A vector Φ belongs to $\mathcal{H}_\pm$ if $S\Phi = \pm\Phi$, where S is the swap operator defined as $S = \sum_{jk} |\varphi_j \otimes \varphi_k\rangle\langle\varphi_k \otimes \varphi_j|$ for arbitrary orthonormal basis $\{\varphi_j\}$ of $\mathcal{H}$. It follows that for all $\varrho, \xi \in \mathcal{S}(\mathcal{H})$

$$S(\varrho \otimes \xi)S^\dagger = \xi \otimes \varrho. \tag{12}$$

Let us denote by $P_\pm$ projections onto symmetric and antisymmetric subspace. Then $S = P_+ - P_-$.

The condition of trivial homogenization implies that

$$U|\psi \otimes \psi\rangle = |\psi \otimes \psi\rangle \tag{13}$$

for all unit vectors $\psi \in \mathcal{H}$. This determines the action of the transformation U on the symmetric subspace $\mathcal{H}_+$. However, it does not give any constraint on the antisymmetric subspace $\mathcal{H}_-$. Since $\mathcal{H} \otimes \mathcal{H} = \mathcal{H}_+ \oplus \mathcal{H}_-$ we can write $U = e^{i\gamma}I_+ \oplus V_-$, where V_- is arbitrary unitary operator on the subspace $\mathcal{H}_-$ and I_+ is the identity operator on the symmetric subspace $\mathcal{H}_+$. In fact, $I_+ = P_+|_{\mathcal{H}_+}$. Any of the collisions of the form $U = e^{i\gamma}I_+ \oplus V_-$ fulfills the conditions of trivial homogenization.

Let us consider the case of $d = \dim\mathcal{H} = 2$ (qubit), for which the antisymmetric space is one dimensional. In one-dimensional Hilbert spaces the unitary operators are complex square roots of the unity, i.e. $V_- = e^{i\beta}I_-$. Explicitly the transformation can be expressed as

$$U|00\rangle = e^{i\gamma}|00\rangle,$$
$$U|11\rangle = e^{i\gamma}|11\rangle,$$
$$U(|01\rangle + |10\rangle) = e^{i\gamma}(|01\rangle + |10\rangle),$$
$$U(|01\rangle - |10\rangle) = e^{i(\gamma+\beta)}(|01\rangle - |10\rangle).$$

Using the identities $e^{i\gamma} = e^{i(\gamma+\beta/2)}e^{-i\beta/2}$, $e^{i(\gamma+\beta)} = e^{i(\gamma+\beta/2)}e^{i\beta/2}$ and discarding the irrelevant global phase factor $e^{i(\gamma+\beta/2)}$ we get a one-parametric set of solutions

$$U_\eta = \cos\eta I + i\sin\eta S, \tag{14}$$

where we set $\eta = -\beta/2$.

In case of more dimensional space (where $d > 2$) the set of unitary transformations is larger. However, when we fix $V_+ = e^{i\beta}I_-$, we get the same one-parametric set of solutions

$$U_\eta = \cos\eta I + i\sin\eta S = e^{i\eta S}. \tag{15}$$

Any element of this class of unitary operators we will call *a partial swap*. In what follows we will focus on the collision dynamics generated by partial swap collisions.

2.2. *Partial swap collisions*

In what follows we denote $\sin\eta = s$ and $\cos\eta = c$. In the process of homogenization, the system particle interacts sequentially with one of the N particles of the reservoir through the transformation $U_\eta = cI + isS$. After the first collision the system and the first reservoir particle undergo the evolution

$$\varrho_S^{(0)} \otimes \xi \mapsto U_\eta(\varrho_S^{(0)} \otimes \xi)U_\eta^\dagger = c^2\varrho_S^{(0)} \otimes \xi + s^2\xi \otimes \varrho_S^{(0)} + ics[S, \varrho_S^{(0)} \otimes \xi].$$

The states of the system particle and of the reservoir particle are obtained as partial traces. Specifically, after the system particle is in the state described by the density operator

$$\varrho_S^{(1)} = c^2\varrho_S^{(0)} + s^2\xi + ics[\xi, \varrho_S^{(0)}], \tag{16}$$

while the first reservoir particle is now in the state

$$\xi_1' = s^2\varrho_S^{(0)} + c^2\xi + ics[\varrho_S^{(0)}, \xi]. \tag{17}$$

When we recursively apply the partial-swap transformation then after the interaction with the n-th reservoir particle, we obtain

$$\varrho_S^{(n)} = c^2\varrho_S^{(n-1)} + s^2\xi + ics[\xi, \varrho_S^{(n-1)}], \tag{18}$$

as the expression for the density operator of the system particle, while the n-th reservoir particle is in the state

$$\xi_n' = s^2\varrho_S^{(n-1)} + c^2\xi + ics[\varrho_S^{(n-1)}, \xi]. \tag{19}$$

2.3. System's convergence

Our next task is to analyze the remaining conditions on the homogenization process. First task is to show that $\varrho_S^{(n)}$ monotonically converges to a δ-vicinity of ξ for all parameters $\eta \neq 0$. We will utilize the *Banach fixed point theorem*[13] claiming that iterations of a *strictly contractive* mapping converge to the single-point contraction. This point is unique and determined by the condition $\mathcal{E}[\xi] = \xi$. A channel $\mathcal{E}$ is called *strictly contractive* if it fulfills the inequality $D(\mathcal{E}[\varrho], \mathcal{E}[\xi]) \leq kD(\varrho, \xi)$ with $0 \leq k < 1$ for all $\varrho, \xi \in \mathcal{S}$.

First of all, it is easy to verify that ξ is a fixed point of $\mathcal{E}_\xi$, where $\mathcal{E}_\xi$ is defined by the Eq.(16). As a distance we will use the Hilbert-Schmidt distance

$$D(\varrho_1, \varrho_2) = ||\varrho_1 - \varrho_2|| = \sqrt{\mathrm{tr}\varrho_1^2 + \mathrm{tr}\varrho_2^2 - 2\mathrm{tr}\varrho_1\varrho_2} \qquad (20)$$

where ϱ_1, ϱ_2 represent quantum states. Consequently,

$$D^2(\mathcal{E}[\varrho_1], \mathcal{E}[\varrho_2]) = c^4 D^2(\varrho_1, \varrho_2) + c^2 s^2 ||[\xi, \varrho_1 - \varrho_2]||^2 \qquad (21)$$

where we used that $\mathcal{E}[\varrho] = c^2\varrho + s^2\xi + ics[\xi, \varrho]$. Defining an operator $\Delta = \varrho_1 - \varrho_2$ our aim is to prove the relation $||[\xi, \Delta]||^2 \leq ||\Delta||^2$. Assuming the spectral decomposition of ξ equals $\xi = \sum_k \lambda_k |k\rangle\langle k|$ and performing the trace in the basis $|k\rangle$ we obtain

$$
\begin{aligned}
||\xi\Delta - \Delta\xi||^2 &= \mathrm{tr}(\xi\Delta - \Delta\xi)^\dagger(\xi\Delta - \Delta\xi) = 2\mathrm{tr}\xi^2\Delta^2 - 2\mathrm{tr}(\xi\Delta)^2 \\
&= 2\sum_j \lambda_j^2 \langle j|\Delta^2|j\rangle - 2\sum_{j,k}\lambda_j\lambda_k|\langle j|\Delta|k\rangle|^2 \\
&= 2\sum_{j,k}\lambda_j^2\langle j|\Delta|k\rangle\langle k|\Delta|j\rangle - 2\sum_{j,k}\lambda_j\lambda_k|\langle j|\Delta|k\rangle|^2 \\
&= \sum_{j,k}(2\lambda_j^2 - 2\lambda_j\lambda_k)|\langle j|\Delta|k\rangle|^2 \\
&= \sum_{j,k}(\lambda_j^2 + \lambda_k^2 - 2\lambda_j\Lambda_k)|\langle j|\Delta|k\rangle|^2 = \sum_{j,k}(\lambda_j - \lambda_k)^2|\langle j|\Delta|k\rangle|^2 \\
&\leq \sum_{j,k}|\langle j|\Delta|k\rangle|^2 = \mathrm{tr}\Delta^\dagger\Delta = ||\Delta||^2
\end{aligned}
$$

where we employed the hermiticity of Δ ($|\langle j|\Delta|k\rangle|^2 = |\langle l|\Delta|j\rangle|^2$) and positivity of ξ which means $|\lambda_1 - \lambda_2| \leq 1$. The proved inequality enables us to write

$$D(\mathcal{E}[\varrho_1], \mathcal{E}[\varrho_2]) \leq |c|D(\varrho_1, \varrho_2) \qquad (22)$$

i.e. the contractivity coefficient k is determined by the parameter $c = \cos\eta$ of the partial swap and the map $\mathcal{E}$ is contractive whenever $|c| < 1$. This result is important because it ensures that in the limit of infinite steps the system will be not only δ-homogenized, but will be described exactly by the state ξ of the reservoir. It means the distance $D(\varrho_S^{(n)}, \xi) \to 0$ vanishes with the number of interactions n.

2.4. Stability of the reservoir

In order to satisfy the last condition (stability of the reservoir) we need to evaluate the distances $D(\xi_j', \xi)$. For each value of η we can specify δ and N_δ. However, the question is whether for arbitrary value of $\delta > 0$ there is a suitable collision U_η and how large is the value of N_δ for which $D(\varrho_S^{(N_\delta)}, \xi) \leq \delta$.

For the system we can combine the contractivity bound $D(\mathcal{E}[\varrho], \mathcal{E}[\xi]) \leq |c| D(\varrho, \xi)$ with the fact that $\mathcal{E}[\xi] = \xi$ to obtain an estimate

$$D(\xi, \varrho_S^{(n)}) \leq |c|^n D(\xi, \varrho_S^{(0)}) = \sqrt{2}|c|^n \leq \delta. \tag{23}$$

Consequently,

$$N_\delta = \frac{\ln(\delta/\sqrt{2})}{\ln|c|}, \tag{24}$$

but the potential values of δ we need to specify from the stability of the reservoir. A direct calculation gives the following bound

$$\begin{aligned}
D(\xi, \xi_n') &= ||\xi - c^2\xi - s^2\varrho_S^{(n-1)} - ics[\varrho_S^{(n-1)}, \xi]|| \\
&\leq s^2||\xi - \varrho_S^{(n-1)}|| + |cs| \cdot ||[\varrho_S^{(n-1)}, \xi]|| \\
&\leq s^2 D(\xi, \varrho_S^{(n-1)}) + 2|cs| \cdot ||\varrho_S^{(n-1)}|| \cdot ||\xi|| \\
&\leq \sqrt{2}|sc|(|s| \cdot |c|^{n-2} + \sqrt{2}) \\
&< \sqrt{2}|sc|(1 + \sqrt{2}) \equiv \delta.
\end{aligned} \tag{25}$$

Important is that δ can be arbitrarily small, hence δ-homogenization is achievable for a restricted class of collisions U_η satisfying the identity $(2 + \sqrt{2})|\sin\eta\cos\eta| \leq \delta$. The number of steps N_δ we achieve if the value of δ is inserted into the formula for N_δ, i.e.

$$N_\delta \approx \frac{\ln\left[(1 + \sqrt{2})|sc|\right]}{\ln|c|}. \tag{26}$$

Let us note that one can derive more precise expression for N_δ and δ, but this is not important for our further purposes.

In summary, the class of partial swap operators U_η satisfies the conditions for δ-homogenization process.

Remark 2.1. *Swap in the classical theory.* Let us note that in the classical picture of thermalization process we also use model of mutual collisions of the system with its thermal reservoir. In these (inelastic) collisions the particles exchange energy and momentum. In some sense, this process can be understood as a transformation that partially "swaps" energy and momentum of involved particles. Thus, from this perspective the homogenizing properties of partial swap collisions are not that surprising and can be viewed as the quantum analogue of classical collisions.

2.5. *Invariance of single-particle average state*

In this section we will allow particles forming the reservoir interacts among each other, but interactions are always pairwise. Partial swap operators possess one unique property, which is important from the point of view of more realistic thermalization process and reservoir's stability. Consider Ω being an n-partite composite system. It can be written in the form

$$\Omega = \varrho_1 \otimes \cdots \otimes \varrho_n + \Gamma, \qquad (27)$$

where Γ is a traceless hermitian operators and $\varrho_j = \mathrm{tr}_{\bar{j}}\Omega$ are the states of individual subsystems. Applying a partial swap collision between jth and kth subsystems we obtain a state

$$\Omega' = c^2\Omega + s^2 S_{jk}\Omega S_{jk} + ics[\Omega, S_{jk}] \qquad (28)$$

$$= \varrho_1' \otimes \cdots \otimes \varrho_n' + \Gamma', \qquad (29)$$

where S_{jk} denotes the swap operator between the subsystems j and k,

$$\varrho_j' = c^2\varrho_j + s^2\varrho_k + ics[\varrho_j, \varrho_k] \qquad (30)$$

$$\varrho_k' = c^2\varrho_k + s^2\varrho_j - ics[\varrho_j, \varrho_k] \qquad (31)$$

$$\varrho_l' = \varrho_l \quad \text{for} \quad l \neq j, k, \qquad (32)$$

and Γ' is the traceless part. It is straightforward to see that

$$\bar{\varrho}_{\mathrm{one}} = \frac{1}{n}\sum_j \varrho_j = \frac{1}{n}\sum_j \varrho_j' = \bar{\varrho}_{\mathrm{one}}', \qquad (33)$$

thus, in the collision model provided by partial swap collisions the *average one-particle state* $\bar{\varrho}_{\mathrm{one}}$ is preserved. If taken the single-particle state as the quantum analogue of temperature, then this property means that partial swap interactions preserves the "quantum" temperature.

3. Quantum decoherence via collisions

In this section we will focus on quantum dynamics of decoherence. Let us note that in literature on quantum information the decoherence is used to name any nonunitary channel and the channels we are interested in are called (generalized) phase-damping channels. In accordance with our previous work[4] we will understand by decoherence a channel with the following properties:

- Preservation of the diagonal elements of a density matrix with respect to a given (decoherence) basis. Let us denote by $\mathcal{B} = \{\varphi_1, \ldots, \varphi_d\}$ the *decoherence basis* of $\mathcal{H}$. Then

$$\mathcal{E} \circ \mathrm{diag}_\mathcal{B} = \mathrm{diag}_\mathcal{B}, \tag{34}$$

  where $\mathrm{diag}_\mathcal{B}$ is a channel diagonalizing density operators in the basis $\mathcal{B}$, i.e. $\varrho \to \mathrm{diag}_\mathcal{B}[\varrho]$.
- Vanishing of the off-diagonal elements of a density matrix with respect to a given (decoherence) basis. In particular, asymptotically the iterations of the decoherence channels results in the diagonalization of the density matrix, i.e.

$$\lim_{n\to\infty} \mathcal{E}^n = \mathrm{diag}_\mathcal{B}. \tag{35}$$

Our tasks is similar as in the previous part: find collisions U generating the decoherence in basis $\mathcal{B}$.

The basis preservation implies that

$$|\varphi_j \otimes \varphi_k\rangle \mapsto e^{i\eta_{jk}} |\varphi_j \otimes \varphi'_{jk}\rangle, \tag{36}$$

where η_{jk} are phases and for each j the unit vectors $\{\varphi'_{jk}\}_k$ form an orthonormal basis of $\mathcal{H}$. It is not difficult to verify that the unitary operators describing the collisions have the following form

$$U = \sum_j |\varphi_j\rangle\langle\varphi_j| \otimes V_j, \tag{37}$$

where $V_j = \sum_k e^{i\eta_{jk}} |\varphi'_{jk}\rangle\langle\varphi_k|$ are unitary operators on individual systems of the environment. Such operators form a class of *controlled unitaries* with the system playing the role of the control system and the environment playing the role of the target system.

A single controlled unitary collision induce the following channel on the system

$$\mathcal{E}[\varrho] = \mathrm{tr}_{\mathrm{env}}[U(\varrho \otimes \xi)U^\dagger] = \sum_{j,k} |\varphi_j\rangle\langle\varphi_j|\varrho|\varphi_k\rangle\langle\varphi_k| \mathrm{tr} V_j \xi V_k^\dagger. \tag{38}$$

Let us note an interesting feature that under the action of this channel the matrix elements of density operators do not mix, i.e. the output value of $\varrho'_{jk} = \langle \varphi_j | \mathcal{E}[\varrho] \varphi_k \rangle$ depends only on the value of $\varrho_{jk} = \langle \varphi_j | \varrho \varphi_k \rangle$. In order to fulfill the second decoherence condition it is sufficient to show that for $j \neq k$

$$|\varrho'_{jk}| < |\varrho_{jk}| \,. \tag{39}$$

For any unitary operator $W = \sum_l e^{iw_l} |l\rangle\langle l|$

$$|\mathrm{tr}\xi W| = |\sum_l e^{iw_l} \langle l|\xi|l\rangle| = |\sum_l p_l e^{iw_l}| \leq 1 \,, \tag{40}$$

where $p_l = \langle l|\xi|l\rangle$ are probabilities. This inequality is saturated only if ξ is an eigenvector of W. It follows that the inequality

$$|\varrho'_{jk}| = |\varrho_{jk}| \cdot |\mathrm{tr}V_j \xi V_k^\dagger| \leq |\varrho_{jk}| \tag{41}$$

holds, hence the off-diagonal elements are non-increasing. They are strictly decreasing if and only if $|\mathrm{tr}V_j \xi V_k^\dagger| < 1$ for all $j \neq k$, which is the case if ξ is not an eigenvector of any of the operators $V_k^\dagger V_j$.

In summary, we have shown that decoherence processes are intimately related with controlled unitary operators. In particular, the decoherence processes are generated by controlled unitary collisions $U = \sum_j |\varphi_j\rangle\langle\varphi_j| \otimes V_j$ satisfying $|\mathrm{tr}V_j \xi V_k^\dagger| < 1$ for all $j \neq k$. In such case

$$\lim_{n\to\infty} \mathcal{E}^n[\varrho] = \mathrm{diag}_\mathcal{B}[\varrho] \,, \tag{42}$$

where $\mathrm{diag}_\mathcal{B}$ is a channel diagonalizing the input density operator in the basis $\mathcal{B}$. Let us note that there are always states ξ (eigenvectors of $V_k^\dagger V_j$) for which the decoherence is not achieved, because for them some of the off-diagonal elements are preserved. On the other side, for each controlled unitary collision U there are always states of the environment ξ inducing the decoherence of all off-diagonal matrix elements.

3.1. *Simultaneous decoherence of the system and the environment*

Could it happen that both the environment and the system are decohering simultaneously? In the same decoherence basis? We will see that answers to both these questions are positive.

For the considered collision model driven by controlled unitary interactions we obtain that the evolution of the particles in the reservoir is described by the following channels

$$\mathcal{N}[\xi] = \mathrm{tr}_{\mathrm{sys}}[U(\varrho \otimes \xi)U^\dagger] = \sum_j \varrho_{jj} V_j \xi V_j^\dagger . \qquad (43)$$

Such channels are known as *random unitary channels.*

It is known that random unitary channels can describe decoherence channels and it was a surprising result that decoherences are not necessarily random unitary channels.[15] We do not need to go into the details of this relation to see that both the system and the environment can decohere. It follows from our previous discussion that environment will decohere if U is controlled unitary transformation of the form

$$U = \sum_j U_j \otimes |\psi_j\rangle\langle\psi_j| , \qquad (44)$$

where $\{\psi_j\}$ is the decoherence basis of the particles in the environment. That is, the question is whether there are unitary operators that can be written as

$$U = \sum_j U_j \otimes |\psi_j\rangle\langle\psi_j| = \sum_j |\varphi_j\rangle\langle\varphi_j| \otimes V_j , \qquad (45)$$

for suitable decoherence bases $\{\varphi_j\}$, $\{\psi_j\}$ and unitary operators U_j, V_j.

Example 3.1. *CTRL-NOT.*

$$U_{\mathrm{ctrl-NOT}} = |0\rangle\langle0| \otimes I + |1\rangle\langle1| \otimes \sigma_x = I \otimes |+\rangle\langle+| + \sigma_z \otimes |-\rangle\langle-| \quad (46)$$
$$= |0+\rangle\langle0+| + |0-\rangle\langle0-| + |1+\rangle\langle1+| - |1-\rangle\langle1-| , \qquad (47)$$

where $|\pm\rangle = (|0\rangle \pm |1\rangle)/\sqrt{2}$.

Proposition 3.1. *A unitary operator U describes a collision with simultaneous decoherence of both interacting systems if and only if $U = \sum_j |\varphi_j\rangle\langle\varphi_j| \otimes V_j$ and the unitary operators V_j commute with each other.*

Proof. By definition, the simultaneous decoherence requires simultaneous preservation of decoherence bases $\{\varphi_j\}$ and $\{\psi_k\}$, i.e.

$$|\varphi_j \otimes \psi_k\rangle \to e^{i\eta_{jk}}|\varphi_j \otimes \psi_k\rangle . \qquad (48)$$

It follows that

$$U = \sum_{jk} e^{i\eta_{jk}}|\varphi_j\rangle\langle\varphi_j| \otimes |\psi_k\rangle\langle\psi_k| , \qquad (49)$$

and we can define

$$U_k = \sum_j e^{i\eta_{jk}}|\varphi_j\rangle\langle\varphi_j|, \qquad V_j = \sum_k e^{i\eta_{jk}}|\psi_k\rangle\langle\psi_k|, \qquad (50)$$

to get the required expression

$$U = \sum_k U_k \otimes |\psi_k\rangle\langle\psi_k| = \sum_j |\varphi_j\rangle\langle\varphi_j| \otimes V_j. \qquad (51)$$

It is straightforward to see that operators $\{V_j\}$ and $\{U_k\}$ form sets of mutually commuting elements.

If the unitary operators V_j commute with each other, then in their spectral form $V_j = \sum_j e^{i\eta_{jk}}|\psi_k\rangle\langle\psi_k|$. Clearly, $U = \sum_j |\varphi_j\rangle\langle\varphi_j| \otimes V_j = \sum_k U_k \otimes |\psi_k\rangle\langle\psi_k|$ with $U_k = \sum_j e^{i\eta_{jk}}|\varphi_j\rangle\langle\varphi_j|$. Therefore, U generates decoherence in both interacting systems.   □

Setting $\psi_k \equiv \varphi_k$ the system's decoherence basis and the environment's decoherence basis coincide. In such case the system and the environment will decohere with respect to the same basis. An example of such interaction for the case of qubit is the CTRL-Z operator

$$U_{\text{ctrl-z}} = |0\rangle\langle0| \otimes I + |1\rangle\langle1| \otimes \sigma_z = I \otimes |0\rangle\langle0| + \sigma_z \otimes |1\rangle\langle1|, \qquad (52)$$

generating decoherence in the computational basis $|0\rangle, |1\rangle$. Moreover, this collision is symmetric with respect to exchange of the role of the control and the target.

4. Entanglement in collision models

The phenomenon of quantum entanglement is considered as one of the key properties of quantum theory. Our aim is not to discuss the philosophical background of this concept,[16,17] but rather focus on the dynamics of entanglement in the considered collision models. We say a state is *separable* if it can be written as a convex combination of factorized states, i.e.

$$\varrho_{\text{sep}} = \sum_j p_j \xi_1^{(j)} \otimes \cdots \otimes \xi_n^{(j)}. \qquad (53)$$

A state is *entangled* if it is not separable. The question whether a given state is entangled, or separable turns out to be very difficult. Partial results are achieved for the system of qubits and therefore in this section we will restrict our discussion only to qubits. However, some of the qualitative properties will be extendible also to larger systems.

A *tangle* is a measure of bipartite entanglement defined as

$$\tau(\omega) = \min_{\omega = \sum_j p_j |\psi_j\rangle\langle\psi_j|} \sum_j p_j \tau(\psi_j), \qquad (54)$$

where $\tau(\psi) = S_{\mathrm{lin}}(\mathrm{tr}_1 |\psi\rangle\langle\psi|) = 2(1 - \mathrm{tr}(\mathrm{tr}_1 |\psi\rangle\langle\psi|)^2) = 4 \det \mathrm{tr}_1[|\psi\rangle\langle\psi|]$ is the *linear entropy* of the state of one of the subsystems. Let us note that for pure bipartite states $S_{\mathrm{lin}}(\mathrm{tr}_1 |\psi\rangle\langle\psi|) = S_{\mathrm{lin}}(\mathrm{tr}_2 |\psi\rangle\langle\psi|)$. Due to seminal paper of Wootters[18] $\tau(\omega) = C^2(\omega)$, where $C(\omega)$ is the so-called *concurrence*

$$C = \max\{0, 2\max\{\sqrt{\lambda_j}\} - \sum_j \sqrt{\lambda_j}\}, \qquad (55)$$

where $\{\lambda_j\}$ are the eigenvalues of $R = \omega(\sigma_y \otimes \sigma_y)\omega^*(\sigma_y \otimes \sigma_y)$.

For multiqubit systems the tangle satisfies the so-called *monogamy relation* (originally conjectures by Coffman et al.[19] and proved by Osborne et al.[20])

$$\sum_{k, k \neq j} \tau_{jk} \leq \tau_j, \qquad (56)$$

where $\tau_j \equiv \tau_{j\bar{j}}$ and $\bar{j}$ denotes the set of all qubits except the jth one, i.e. τ_j is a tangle between the jth qubit and rest of the qubits considered as a single system. The above inequality is called *CKW inequality*. If the multiqubit system is in a pure state Ψ then

$$\tau_j = 4 \det \mathrm{tr}_{\bar{j}}[|\Psi\rangle\langle\Psi|]. \qquad (57)$$

Example 4.1. *Saturation of CKW inequalities.* At the end of the original paper[19] it is stated that states in the subspace covered by the basis $\{|1\rangle_j \otimes |0^{\otimes N-1}\rangle_{\bar{j}}\}$ saturates the CKW inequalities. In what follows we will show that any state of the form

$$|\Psi\rangle = \alpha_0 |0^{\otimes N}\rangle + \sum_{j=1}^{N} \alpha_j |1\rangle_j \otimes |0^{\otimes(N-1)}\rangle \qquad (58)$$

saturates the CKW inequalities.

The reduced bipartite density matrices are of the form

$$\varrho = \begin{pmatrix} a & d & e & 0 \\ d^* & b & f & 0 \\ e^* & f^* & c & 0 \\ 0 & 0 & 0 & 0 \end{pmatrix} \qquad (59)$$

and only one element determines the tangle of such state, namely

$$\tau(\varrho) = 4 f f^*. \qquad (60)$$

It follows that only the matrix element standing with the $|01\rangle\langle 10|$ term will be important for us. Direct calculations lead us to value $f = \alpha_j \alpha_k^*$ for the pair of jth and kth, hence

$$\tau_{jk} = 4|\alpha_j|^2 |\alpha_k|^2 . \tag{61}$$

In the next step we evaluate the tangle between the j-th particle and the rest of the system. The state of single particle is described by matrix

$$\varrho_j = \begin{pmatrix} |\alpha_0|^2 + \sum_{k \neq j} |\alpha_k|^2 & \alpha_0 \alpha_j^* \\ \alpha_j \alpha_0^* & |\alpha_j|^2 \end{pmatrix} \tag{62}$$

Now it is easy to check that

$$\tau_j = 4 \det \varrho_j = |\alpha_j|^2 \sum_{k \neq j} |\alpha_k|^2 = \sum_{k \neq j} \tau_{jk} \tag{63}$$

and therefore the CKW inequalities are saturated, i.e. $\Delta_j = \tau_j - \sum_{k \neq j} \tau_{jk} = 0$ for all j.

4.1. *Entanglement in partial swap collision model*

A general pure input state equals

$$|\Psi\rangle = |\psi\rangle_S \otimes |\varphi\rangle_1 \otimes \cdots \otimes |\varphi\rangle_N . \tag{64}$$

Since $[I, V \otimes V] = [S, V \otimes V] = 0$ for all unitary operators V on $\mathcal{H}$ it follows that also $[U_n, V \otimes V] = 0$ for all unitary V. Moreover, the value of entanglement is invariant under local unitary transformations. Therefore, without loss of generality we can set $|\varphi\rangle = |0\rangle$ and $|\psi\rangle = \alpha|0\rangle + \beta|1\rangle$. After n interactions

$$|\Psi_n\rangle = \alpha|0^{\otimes(N+1)}\rangle + \beta c^n |1\rangle_S \otimes |0^{\otimes N}\rangle + \beta \sum_{l=1}^{n} |1\rangle_l \otimes |0^{\otimes N_{\bar{l}}}\rangle \left[isc^{l-1} e^{i\eta(n-l)} \right],$$

where $N_{\bar{l}}$ denotes a system of N qubits obtained by replacing the lth qubit from the reservoir by the system qubit.

Since $|\Psi_n\rangle$ belongs to the class of states discussed in Example 4.1, we know that the state saturates CKW inequalities. We can used the derived formulas to write

$$\tau_{jk}(n) = 4|\beta|^4 s^4 c^{2(j+k-2)} ; \tag{65}$$

$$\tau_{0k}(n) = 4|\beta|^4 s^2 c^{2(n+k-1)} ; \tag{66}$$

$$\tau_j(n) = 4|\beta|^4 s^2 c^{2(j-1)} (1 - s^2 c^{2(j-1)}) ; \tag{67}$$

$$\tau_0(n) = 4|\beta|^4 c^{2n} (1 - c^{2n}) . \tag{68}$$

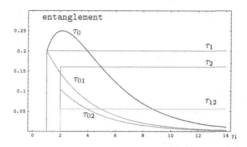

Figure 2.   Entanglement in the homogenization process.

Let us note that these formulas are valid only if $j, k \leq n$. Otherwise the quantities vanish.

These results show that the system particle acts as a mediator entangling the reservoir particles which have never interacted directly. It is obvious that the later the two reservoir qubits interact with the system particle, the smaller the degree of their mutual entanglement is. Nevertheless, this value remains constant and does not depend on the subsequent evolution of the collision model (i.e., it does not depend on the number of interactions n). On the other hand the entanglement between the system particle and kth reservoir particle (k arbitrary) increases at the moment of interaction and then monotonously decreases with the number of interaction steps.

4.2. *Entanglement in controlled unitary collision model*

As before, let us assume that initially the system and the reservoir of N qubits are described by a pure input state

$$|\Omega\rangle = |\psi\rangle_S \otimes |\varphi\rangle_1 \otimes \cdots \otimes |\varphi\rangle_N \,, \qquad (69)$$

with $\psi = \alpha|0\rangle + \beta|1\rangle$. After n controlled unitary collisions we get

$$|\Omega_n\rangle = [\alpha|0\rangle \otimes |\varphi_0^{\otimes n}\rangle + \beta|1\rangle \otimes |\varphi_1^{\otimes n}\rangle] \otimes |\varphi^{\otimes(N-n)}\rangle \,, \qquad (70)$$

where we set $|\varphi_0\rangle = V_0|\varphi\rangle$ and $|\varphi_1\rangle = V_1|\varphi\rangle$.

Tracing out the system we end up with the reservoir described by the state

$$\omega_{\text{env}}(n) = (|\alpha|^2|\varphi_0^{\otimes n}\rangle\langle\varphi_0^{\otimes n}| + |\beta|^2|\varphi_1^{\otimes n}\rangle\langle\varphi_1^{\otimes n}|) \otimes |\varphi\rangle\langle\varphi|^{\otimes(N-n)} \,, \qquad (71)$$

which is clearly separable, i.e. $\tau_{jk}(n) = 0$. In this case no entanglement is created within the environment.

A state of the system qubit and kth qubit from the reservoir equals (providing that $n \geq k$)

$$\varrho_{0k}(n) = |\alpha|^2 |0\varphi_0\rangle\langle 0\varphi_0| + |\beta|^2 |1\varphi_1\rangle\langle 1\varphi_1|$$
$$+ \alpha\beta^* |\langle\varphi_0|\varphi_1\rangle|^{(n-1)} |0\varphi_0\rangle\langle 1\varphi_1| + c.c\,, \qquad (72)$$

thus for the tangle we have

$$\tau_{0k}(n) = 4|\alpha|^2|\beta|^2 |\langle\varphi_0|\varphi_1\rangle|^{2(n-1)} |\langle\varphi_0|\varphi_1^\perp\rangle|^2\,. \qquad (73)$$

If $n < k$ then $\tau_{0k}(n) = 0$. Further,

$$\varrho_0(n) = |\alpha|^2 |0\rangle\langle 0| + |\beta|^2 |1\rangle\langle 1| + \alpha\beta^* |\langle\varphi_0|\varphi_1\rangle|^n |0\rangle\langle 1| + c.c\,, \qquad (74)$$
$$\varrho_k(n) = |\alpha|^2 |\varphi_0\rangle\langle\varphi_0| + |\beta|^2 |\varphi_1\rangle\langle\varphi_1|\,, \qquad (75)$$
$$\qquad (76)$$

resulting in the following values for tangle

$$\tau_0(n) = 4|\alpha|^2|\beta|^2 (1 - |\langle\varphi_0|\varphi_1\rangle|^{2n})\,, \qquad (77)$$
$$\tau_k(n) = 4|\alpha|^2|\beta|^2 |\langle\varphi_0|\varphi_1^\perp\rangle|^2\,. \qquad (78)$$

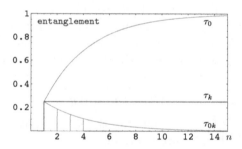

Figure 3.   Entanglement in the decoherence process.

In summary, during the decoherence collision model the qubits in the reservoir are not entangled at all. Interaction of the system qubit with an individual reservoir's qubit entangles these pair of qubits. However, with the number of collision this bipartite entanglement vanishes and finally $\tau_{0k} = 0$ for all reservoir's qubits k. The entanglement between a fixed qubit of the reservoir and rest of the qubits preserves its value it gathered after the interaction. The entanglement between a system qubit and the reservoir is increasing with the number of collisions and approaches the maximal value $\tau_0(\infty) = 4|\alpha|^2|\beta|^2$.

5. Master equations for collision models

Within the collision model the system undergoes a discrete dynamics

$$\varrho \to \mathcal{E}_1[\varrho] \to \mathcal{E}_2[\varrho] \to \cdots \to \mathcal{E}_n[\varrho] \to \cdots , \tag{79}$$

where $\mathcal{E}_k = \mathcal{E} \circ \cdots \circ \mathcal{E} = \mathcal{E}^k$. Let us define $\mathcal{E}_0$ as the identity map $\mathcal{I}$. It is straightforward to see that the sequence of channels $\mathcal{E}_0, \mathcal{E}_1, \mathcal{E}_2, \ldots$ form a discrete semigroup satisfying the relation

$$\mathcal{E}_n \circ \mathcal{E}_m = \mathcal{E}_{n+m} \tag{80}$$

for all $n, m = 0, 1, 2, \ldots$. In this section we will investigate the continuous interpolations of such discrete semigroups. Interpreting the continuous parameter as time we will derive first order differential master equation generating the continuous sequence of linear maps. Our derivation of the master equation will be purely heuristic. We will simply replace n in $\mathcal{E}_n$ by a continuous time parameter t and after that we will have a look what are the consequences.

Let $\varrho_t = \mathcal{E}_t[\varrho_0]$. The time derivative of the state dynamics results in the following differential equation

$$\frac{d\varrho_t}{dt} = \left(\frac{d\mathcal{E}_t}{dt} \right)[\varrho_0] = \left(\frac{d\mathcal{E}_t}{dt} \circ \mathcal{E}_t^{-1} \right)[\varrho_t] = \mathcal{G}_t[\varrho_t] , \tag{81}$$

where $\mathcal{G}_t$ is the generator of the dynamics $\mathcal{E}_t$. Let us note that this approach is very formal. There is no guarantee that the mapping $\mathcal{E}_t$ is a valid quantum channel for each value of t and that the set $\mathcal{E}_t$ is indeed continuous. Another problematic issue is the assumed existence of the inverse of the map $\mathcal{E}_t^{-1}$ for all $t \geq 0$. All these properties need to be checked before we give some credit to the derived master equation.

5.1. *One-parametric semigroups*

In a special case when $\mathcal{G}_t$ is time-independent the solution of the differential equation

$$\frac{d\varrho_t}{dt} = \mathcal{G}[\varrho_t] , \tag{82}$$

with the initial condition $\varrho_{t=0} = \varrho_0$ can be written in the form

$$\varrho_t = e^{\mathcal{G}t}[\varrho_0] ; \quad \mathcal{E}_t = e^{\mathcal{G}t} . \tag{83}$$

Since $e^{\mathcal{G}(t+s)} = e^{\mathcal{G}s} e^{\mathcal{G}t}$ it follows that

$$\mathcal{E}_t \circ \mathcal{E}_s = \mathcal{E}_{t+s} \tag{84}$$

for all $t, s \geq 0$ and $\mathcal{E}_0 = \mathcal{I}$. In such case the continuous set of linear maps $\{\mathcal{E}_t\}$ form *a one-parametric semigroup*, which are playing important role in the theory of open system dynamics. Semigroups of channels are often used to approximate real open system's evolutions. In particular, they are satisfying the so-called *Markovianity condition*[21] restricting the memory effects of the environment. The following theorem specifies the properties of the generator $\mathcal{G}$ assuring the validity of the quantum channel constraints on $\mathcal{E}_t = e^{t\mathcal{G}}$.

Theorem 5.1. *(Lindblad*[22]*)* $\mathcal{E}_t = e^{\mathcal{G}t}$ *are valid quantum channels for all* $t \geq 0$ *if and only if* $\mathcal{G}$ *can be written in the form*

$$\mathcal{G}[\varrho] = -\frac{i}{\hbar}[H, \varrho] + \frac{1}{2}\sum_{jk} c_{jk}([\Lambda_j\varrho, \Lambda_k] + [\Lambda_j, \varrho\Lambda_k]), \qquad (85)$$

where $\{I, \Lambda_1, \ldots, \Lambda_{d^2-1}\}$ *is an orthonormal operator basis consisting of hermitian traceless operators* Λ_j, *i.e.* $\mathrm{tr}\Lambda_j\Lambda_k = \delta_{jk}$, *the coefficients* c_{jk} *form a positive matrix and* H *is a hermitian traceless operator on* $\mathcal{H}$.

5.2. *Divisibility of channels*

Any collision U induced some channel $\mathcal{E}$. Our interpolation should smoothly connect this channel with the identity map $\mathcal{I}$ such that the connecting line is inside the set of channels. Since the set of channels is convex any two points are connected by a line containing only proper quantum channels, i.e. interpolation is surely possible although its continuity is still not guaranteed. Let us have a look on the possibility to interpolate $\mathcal{I}$ and $\mathcal{E}$ such that $\mathcal{E} = e^{\tau\mathcal{G}}$ for some time τ and time-independent Lindblad generator. Without loss of generality we can set $\tau = 1$, i.e. $\mathcal{E} = e^{\mathcal{G}}$. The semigroup property implies that $\mathcal{E} = \mathcal{E}_\varepsilon \circ \mathcal{E}_{1-\varepsilon}$ for any $0 < \epsilon < 1$, or even stronger $\mathcal{E} = \mathcal{E}_\epsilon^{1/\epsilon}$.

It is a surprising fact[23] that there are channels $\mathcal{E}$, which cannot be expressed as a nontrivial composition of some other channels $\mathcal{E}_1, \mathcal{E}_2$. Consequently not all channels can be part of some one-parametric semigroup. Before we will show one example, let us formally define the indivisibility.

Definition 5.1. We say a channel $\mathcal{E}$ is *indivisible* if $\mathcal{E} = \mathcal{E}_1 \circ \mathcal{E}_2$ implies that either $\mathcal{E}_1$, or $\mathcal{E}_2$ are unitary channels (such decomposition is called trivial).

Let us note that one of seemingly counter-intuitive consequences is that unitary channels are indivisible. In a suitable representation the composition of channels is just a product of matrices. Thus for determinants

$$\det(\mathcal{E}_1 \circ \mathcal{E}_2) = (\det \mathcal{E}_1)(\det \mathcal{E}_2) \, . \tag{86}$$

Let us denote by $\mathcal{E}_{\min}$ a channel minimizing the value of determinant, i.e. $\mathcal{E}_{\min} = \arg\min_{\mathcal{E}} \det \mathcal{E}$. Assume $\mathcal{E}_{\min} = \mathcal{E}_1 \circ \mathcal{E}_2$. Since the value of $\det \mathcal{E}_{\min}$ is negative and for channels $|\det \mathcal{E}| \leq 1$, it follows that either $\det \mathcal{E}_1 = 1, \det \mathcal{E}_2 = \det \mathcal{E}_{\min}$, or $\det \mathcal{E}_2 = 1, \det \mathcal{E}_1 = \det \mathcal{E}_{\min}$. But $\det \mathcal{E} = 1$ implies $\mathcal{E}$ is a unitary channel, which proves the triviality of any decomposition of the minimum determinant channel. If $d = 2$ (qubit), then one of the minimum determinant channels is the *optimal universal NOT*,[23] i.e.

$$\mathcal{E}_{\min}[\varrho] = \frac{1}{3}(I + \varrho^T) \, , \tag{87}$$

for which $\det \mathcal{E}_{\min} = -1/27$. The mapping $\varrho \to \varrho^T$ known as the *universal NOT* is not a proper quantum channel, because it is not completely positive and, thus, unphysical. Let us note that channels of minimal determinant are not the only indivisible channels. As far as the authors know a complete characterization of indivisible channels is not known.

For our purposes the introduced concept of indivisibility is not sufficient as we are interested in the existence of continuous semigroup between $\mathcal{I}$ and $\mathcal{E}$ generated by a unitary collision U. We need $\mathcal{E}$ to be *infinitely divisible*, i.e. $\mathcal{E} = \mathcal{E}_{1/n}^n$ for all $n > 0$. It was shown by Denisov[24] that infinitely divisible channel can be expressed in the form $\mathcal{E} = \mathcal{F} \circ e^{\mathcal{G}}$, where $\mathcal{F}$ is an idempotent channel ($\mathcal{F}^2 = \mathcal{F}$) satisfying $\mathcal{F}\mathcal{G} = \mathcal{F}\mathcal{G}\mathcal{F}$ and $\mathcal{G}$ is a Lindblad generator. We are interested in cases when $\mathcal{F} = \mathcal{I}$, thus, $\mathcal{E} = e^{\mathcal{G}}$. Then it follows that the sequence of channels $\mathcal{E}^n = e^{n\mathcal{G}}$ determined by the collision model belongs to the one-parametric semigroup $e^{t\mathcal{G}}$. In other words a discrete semigroup is interpolated by a continuous one. We will see that this is the case for the collisions considered in the homogenization and decoherence processes.

For a given channel induced by a single collision question of interest is whether $\mathcal{E} = e^{\mathcal{G}}$ for some Lindblad generator $\mathcal{G}$. In other words whether $\mathcal{G} = \log \mathcal{E}$ is a valid generator of complete positive dynamics. Since channels $\mathcal{E}$ are not necessarily diagonalizable, the notion of the matrix/operator logarithm is not completely trivial. The logarithm is not unique and any operator X such that $e^X = \mathcal{E}$ is called $\log \mathcal{E}$. We need to search all logarithms in order to find a valid generator of the Lindblad form. Evaluating the logarithm will give us a time-independent generator (if it exists) interpolating the discrete semigroup $\mathcal{E}_n = \mathcal{E}^n$. For both considered collision models the evaluation of the logarithm is not difficult. However, for illustrative purposes in our analysis of homogenization we will follow the *ad hoc* procedure exploiting

(81), which (if successful) could capture also time-dependent generators. For the case of decoherence we will evaluate the logarithm of the induced channel.

5.3. *Bloch sphere parametrization*

The set of operators form a complex linear space endowed with a *Hilbert-Schmidt scalar product* $(X, Y) = \mathrm{tr}X^\dagger Y$. Let $\varphi_1, \ldots, \varphi_d$ be an orthonormal basis of $\mathcal{H}$. Then the set of d^2 operators $\{E_{jk} = |\varphi_j\rangle\langle\varphi_k|\}$ form an orthonormal operator basis, i.e. $\mathrm{tr}E_{j'k'}^\dagger E_{jk} = \langle\varphi_{j'}|\varphi_j\rangle\langle\varphi_k|\varphi_{k'}\rangle = \delta_{jj'}\delta_{kk'}$.

In what follows we will restrict to the case of two-dimensional Hilbert spaces $\mathcal{H}$ (qubits). This will be perfectly sufficient for our later purposes, because we will focus on qubit collision models of homogenization and decoherence. However, many of the properties and procedures can be easily extended to more dimensional case.

Let us start with a very convenient operator basis of qubit systems — the set of *Pauli operators*

$$I = |0\rangle\langle0| + |1\rangle\langle1|, \qquad \sigma_x = |0\rangle\langle1| + |1\rangle\langle0|,$$
$$\sigma_y = -i(|0\rangle\langle1| - |1\rangle\langle0|), \qquad \sigma_z = |0\rangle\langle0| - |1\rangle\langle1|,$$

where $\{|0\rangle, |1\rangle\}$ is an orthonormal basis of $\mathcal{H}$. These operators are hermitian, $\sigma_j = \sigma_j^\dagger$, and mutually orthogonal, i.e. $\mathrm{tr}\sigma_j\sigma_k = 2\delta_{jk}$ (we set $\sigma_0 = I$). They are also unitary, but this property will not be very important for us.

Using Pauli operators the density operators takes the form

$$\varrho = \frac{1}{2}(I + \vec{r} \cdot \vec{\sigma}),\tag{88}$$

where $\vec{\sigma} = (\sigma_x, \sigma_y, \sigma_z)$ and $\vec{r} = \mathrm{tr}\varrho\vec{\sigma}$ is the so-called *Bloch vector*. The positivity constraint restricts its length $|\vec{r}| \leq 1$, i.e. in the Bloch vector parametrization the qubit's states form a unit sphere called *Bloch sphere*. Let us note that such simple picture of state space does not hold for more dimensional quantum systems.

A quantum channel $\mathcal{E}$ is acting on Pauli operators as follows

$$\varrho \to \varrho' = \mathcal{E}[\varrho] = \frac{1}{2}(\mathcal{E}[I] + x\mathcal{E}[\sigma_x] + y\mathcal{E}[\sigma_y] + z\mathcal{E}[\sigma_z]).\tag{89}$$

Using $\mathcal{E}[\sigma_j] = \sum_k \mathcal{E}_{kj}\sigma_k$ with $\mathcal{E}_{kj} = \frac{1}{2}\mathrm{tr}\sigma_k\mathcal{E}[\sigma_j]$. The tracepreservity ensures that $\mathcal{E}_{00} = \frac{1}{2}\mathrm{tr}\mathcal{E}[I] = 1$ and $\mathcal{E}_{0j} = \mathrm{tr}\mathcal{E}[\sigma_j] = 0$ for $j = x, y, z$. That is,

$$\varrho' = \frac{1}{2}\left[I + \sum_{k=x,y,x}\left(\mathcal{E}_{k0} + \sum_{j=x,y,x}\mathcal{E}_{kj}r_j\right)\sigma_k\right].\tag{90}$$

Comparing with the expression $\varrho' = \frac{1}{2}(I + \vec{r}' \cdot \vec{\sigma})$ we get that under the action of a channel $\mathcal{E}$ the Bloch vectors are transformed by an affine transformation

$$\vec{r} \to \vec{r}' = \vec{t} + T\vec{r}, \tag{91}$$

where $t_j = \mathcal{E}_{j0}$ and T is a 3x3 matrix with entries $T_{jk} = \mathcal{E}_{jk}$ for $j, k = x, y, z$. The Lindblad generator takes the form of 4×4 matrix

$$\mathcal{G} = \begin{pmatrix} 0 & 0 & 0 & 0 \\ g_{10} & g_{11} & g_{12} & g_{13} \\ g_{20} & g_{21} & g_{22} & g_{23} \\ g_{30} & g_{31} & g_{32} & g_{33} \end{pmatrix}, \tag{92}$$

where $g_{jk} = \frac{1}{2}\mathrm{tr}\sigma_j\mathcal{G}[\sigma_k]$. Inserting the Lindblad operator-sum form

$$\mathcal{G}[X] = -\frac{i}{\hbar} \sum_{j=x,y,z} h_j[\sigma_j, X] + \frac{1}{2} \sum_{j,k=x,y,z} c_{jk}([\sigma_j, X\sigma_k] + [\sigma_j X, \sigma_k]). \tag{93}$$

into the matrix expression we obtain

$$g_{jk} = 2\sum_l \epsilon_{jkl}h_l + \frac{1}{2}(c_{kj} + c_{jk}) - \sum_l c_{ll}\delta_{jk}, \tag{94}$$

$$g_{k0} = i\sum_{jl} \epsilon_{jlk}c_{jl}. \tag{95}$$

The inverse relations express the parameters c_{jk} and h_j via the elements of the matrix $\mathcal{G}$

$$h_1 = \tfrac{g_{32}-g_{23}}{4}, \quad h_2 = \tfrac{g_{13}-g_{31}}{4}, \quad h_3 = \tfrac{g_{21}-g_{12}}{4}$$

$$c_{jj} = g_{jj} - \tfrac{1}{2}\sum_k g_{kk} \tag{96}$$

$$c_{12} = \tfrac{1}{2}(g_{12} + g_{21} - ig_{30}), \quad c_{21} = \tfrac{1}{2}(g_{12} + g_{21} + ig_{30}),$$

$$c_{23} = \tfrac{1}{2}(g_{23} + g_{32} - ig_{10}), \quad c_{32} = \tfrac{1}{2}(g_{23} + g_{32} + ig_{10}),$$

$$c_{13} = \tfrac{1}{2}(g_{13} + g_{31} + ig_{20}), \quad c_{31} = \tfrac{1}{2}(g_{13} + g_{31} - ig_{20}).$$

5.4. *Master equation for homogenization collision model*

The collision model of quantum homogenization driven by partial swaps results in the sequence of channels being powers of a channel $\mathcal{E}$ defined in Eq.(16), i.e.

$$\vec{r} \to \vec{r}' = c^2\vec{r} + s^2\vec{w} - cs\vec{w} \times \vec{r}, \tag{97}$$

where $\vec{w}$ is the Bloch vector associated with the state ξ and $c = \cos\eta$, $s = \sin\eta$. Choosing a new operator basis $S_j = V\sigma_j V^\dagger$ such that $\xi = \frac{1}{2}(I + wS_3)$

we get

$$
\mathcal{E} = \begin{pmatrix} 1 & 0 & 0 & 0 \\ 0 & c^2 & csw & 0 \\ 0 & -csw & c^2 & 0 \\ s^2w & 0 & 0 & c^2 \end{pmatrix}. \tag{98}
$$

Let us note that the basis transformation $\sigma_j \rightarrow S_j$ corresponds to a rotation of the coordinate system of the Bloch sphere representation. Moreover, we used that the partial swap commutes with unitaries of the form $V \otimes V$, hence the form of the induced map (97) is unaffected, only the vectors are expressed with respect to different coordinate system.

Let us introduce an angle $\theta = \arctan(ws/c) = \arctan(w \tan \eta)$ and a parameter $q = \sqrt{c^2 + w^2 s^2}$. Using the identity $\cos \arctan(x) = 1/\sqrt{1 + x^2}$ we get $q \cos \theta = \cos \eta = c$ and $q \sin \theta = w \sin \eta$, thus

$$
\mathcal{E} = \begin{pmatrix} 1 & 0 & 0 & 0 \\ 0 & cq \cos \theta & cq \sin \theta & 0 \\ 0 & -cq \sin \theta & cq \cos \theta & 0 \\ s^2w & 0 & 0 & c^2 \end{pmatrix}. \tag{99}
$$

The reason for such parametrization becomes clear if we evaluate powers of $\mathcal{E}$

$$
\mathcal{E}^n = \begin{pmatrix} 1 & 0 & 0 & 0 \\ 0 & c^n q^n \cos n\theta & c^n q^n \sin n\theta & 0 \\ 0 & -c^n q^n \sin n\theta & c^n q^n \cos n\theta & 0 \\ w(1 - c^{2n}) & 0 & 0 & c^{2n} \end{pmatrix}.
$$

In the next step we make *ad hoc* assumption and replace n by t/τ, where τ is some time scale and $t \geq 0$ is a continuous time parameter. Introducing the parameters

$$
\Omega = \theta/\tau, \quad c^{2t/\tau} = e^{-\Gamma_1 t}, \quad (cq)^{t/\tau} = e^{-\Gamma_2 t}, \tag{100}
$$

we end up with the continuous set of trace-preserving linear maps

$$
\mathcal{E}_t = \begin{pmatrix} 1 & 0 & 0 & 0 \\ 0 & e^{-\Gamma_2 t} \cos \Omega t & e^{-\Gamma_2 t} \sin \Omega t & 0 \\ 0 & -e^{-\Gamma_2 t} \sin \Omega t & e^{-\Gamma_2 t} \cos \Omega t & 0 \\ w(1 - e^{-\Gamma_1 t}) & 0 & 0 & e^{-\Gamma_1 t} \end{pmatrix}. \tag{101}
$$

It is not difficult to see that they form a one-parametric semigroup ($\mathcal{E}_t \circ \mathcal{E}_s = \mathcal{E}_{t+s}$), but one needs to verify whether for each t these maps are completely positive. Therefore, we calculate the generator $\mathcal{G}$ and verify its properties. In particular, we obtain a time-independent generator

$$\mathcal{G} = \begin{pmatrix} 0 & 0 & 0 & 0 \\ 0 & -\Gamma_2 & -\Omega & 0 \\ 0 & \Omega & -\Gamma_2 & 0 \\ w\Gamma_1 & 0 & 0 & -\Gamma_1 \end{pmatrix} .$$

Using the Eqs.(96) we can verify that the parameters $h_1 = h_2 = 0, h_3 = -\Omega/2$ are real and that the matrix

$$C = \begin{pmatrix} -\Gamma_1/2 & -i2w\Gamma_1 & 0 \\ i2w\Gamma_1 & -\Gamma_1/2 & 0 \\ 0 & 0 & -(\Gamma_1 + 2\Gamma_2)/2 \end{pmatrix} \tag{102}$$

is positive, hence, the generator $\mathcal{G}$ is of Lindblad form. In the operator-sum (Kraus) form the master equation reads

$$\frac{d\varrho}{dt} = i\frac{\Omega}{2\hbar}[S_z, \varrho] - iw\Gamma_1(\Phi_{xy}[\varrho] + \Phi_{yx}[\varrho]) \tag{103}$$

$$- \frac{\Gamma_1}{4}(\Phi_{xx}[\varrho] + \Phi_{yy}[\varrho]) - \frac{\Gamma_1 + 2\Gamma_2}{4}\Phi_{zz}[\varrho] ,$$

where $\Phi_{jk}[\varrho] = \frac{1}{2}([\sigma_j\varrho, \sigma_k] + [\sigma_j, \varrho\sigma_k])$. After a little algebra we obtain

$$\frac{d\varrho}{dt} = i\frac{\Omega}{2\hbar}[S_z, \varrho] - iw\Gamma_1(S_x\varrho S_y - S_y\varrho S_x + i\varrho S_z + iS_z\varrho) \tag{104}$$

$$+ \frac{1}{4}\Gamma_1(S_x\varrho S_x + S_y\varrho S_y - 2\varrho) + \frac{1}{4}(2\Gamma_2 - \Gamma_1)(S_z\varrho S_z - \varrho) .$$

Let us note that for the special choice of parameters the above master equation coincide with the master equation used to model the spontaneous decay of a two-level atom. In particular, setting $\Gamma_1 = 2\Gamma_2 = 2\gamma = -\frac{2}{\tau}\ln\cos\eta$ and ξ being a pure state ($w = 1$) we get

$$\frac{d}{dt}\varrho = -i\frac{\Omega}{2\hbar}[S_z, \varrho] + \frac{\gamma}{2}[2S_-\varrho S_+ - S_-S_+\varrho - \varrho S_+S_-) \tag{105}$$

where we used $S_\pm = (S_x \pm iS_y)/2$.

5.5. *Master equation for decoherence collision model*

In this section we will repeat the same steps as in the previous one, but for collisions described by controlled unitary interactions. Without loss of generality we will assume that the decoherence basis coincide with the eigenvectors of $S_z = V\sigma_z V^\dagger$ operator. The channel induced by a collision $U = |0\rangle\langle 0| \otimes V_0 + |1\rangle\langle 1| \otimes V_1$ reads

$$
\mathcal{E} = \begin{pmatrix} 1 & 0 & 0 & 0 \\ 0 & \lambda\cos\varphi & \lambda\sin\varphi & 0 \\ 0 & -\lambda\sin\varphi & \lambda\cos\varphi & 0 \\ 0 & 0 & 0 & 1 \end{pmatrix} , \tag{106}
$$

where $\mathrm{tr}V_1^\dagger V_0\xi = \lambda e^{i\varphi}$. We can follow the same derivation of master equation as in the case of homogenization.[4] However, for the illustration purposes we will describe the second procedure based on evaluation of $\log\mathcal{E}$. Unlike the case of homogenization the decoherence channel $\mathcal{E}$ defines a hermitian matrix, thus, the logarithm is pretty easy to calculate exploiting the simple functional calculus for hermitian operators.

Let us start with the observation that

$$
\lambda(\cos\varphi I + i\sin\varphi\sigma_y) = e^{\ln\lambda I} e^{i(\varphi+2k\pi)\sigma_y} = e^{\ln\lambda I + i(\varphi+2k\pi)\sigma_y} , \tag{107}
$$

where k is arbitrary natural number. For $k = 0$ the logarithm is called principal. For our purposes it is sufficient to consider only this one, because we are restricted to angles inside the interval $[0, 2\pi]$. Since this matrix is the central part of the matrix $\mathcal{E}$, it follows that

$$
\mathcal{G} = \log\mathcal{E} = \begin{pmatrix} 0 & 0 & 0 & 0 \\ 0 & \ln\lambda & -\varphi & 0 \\ 0 & \varphi & \ln\lambda & 0 \\ 0 & 0 & 0 & 0 \end{pmatrix} , \tag{108}
$$

is the generator of the semigroup dynamics containing the channel $\mathcal{E}$. Is $\mathcal{E}_t = e^{\mathcal{G}t}$ a valid quantum channel for any t?

Using the relations specified in Eqs.(96) we found that the only nonzero parameters are

$$
h_3 = \varphi/2 , \qquad c_{33} = -\ln\lambda/2 . \tag{109}
$$

Since $c_{33} \geq 0$ it follows that the matrix C is positive. Therefore

$$
\begin{aligned}
\mathcal{G}[\varrho] &= -i\frac{\varphi}{2\hbar}[S_z, \varrho] - \frac{\ln\lambda}{4}([S_z, \varrho S_z] + [S_z\varrho, S_z]) \\
&= -i\frac{\varphi}{2\hbar}[S_z, \varrho] - \frac{\ln\lambda}{2}(S_z\varrho S_z - \varrho) ,
\end{aligned}
$$

defines a correct Lindblad generator. Using $H = \frac{1}{2}\varphi S_z$ and $\gamma = (2\ln\lambda)/\varphi^2$ we obtain a well-known master equation

$$
\frac{d\varrho}{dt} = -\frac{i}{\hbar}[H, \varrho] - \frac{\gamma}{2}[H, [H, \varrho]] , \tag{110}
$$

which is used to model the decoherence also for more dimensional systems.

6. Conclusions

In these lectures we introduced and investigated the simple collision model to capture relevant features and properties of quantum open system dynamics. We focused on two particular collision models determined by the choice of the unitary transformations describing the individual collisions:

- *Homogenization* induced by the partial swap interactions $U_\eta = \cos \eta I + i \sin \eta S$. This process was motivated by the classical thermalization process (0th law of thermodynamics).
- *Decoherence* induced by the controlled unitary interactions $U = \sum_j |\varphi_j\rangle\langle\varphi_j| \otimes V_j$. This process describes the disappearance of the quantumness of quantum systems.

It is known that creation of entanglement requires interactions between quantum systems. In our collision model we start from completely factorized state. It is an interesting question what type of multipartite entanglement is created via a well-defined sequence of bipartite collisions. We haven't provided a definite answer to this problem, but we illustrated that the created entanglement of the particular collision models is not trivial. In particular, the partial swap interaction creates W-type of entanglement for which all the involved systems are pairwisely entangled although they did not interact directly. Moreover, during the whole evolution the CKW inequalities are saturated. The entanglement created in the decoherence collision model is of completely different quality. In this case the created entanglement is of GHZ-type meaning that pairwise entanglement in the reservoir vanishes, however the system is still entangled. The reservoir itself (the system is traced out) is in a separable state. The collision models can be understood as a (simple) preparation processes aiming to create multipartite entanglement. And it is of interest to understand what quality and quantity of entanglement it is capable of.

The well-defined sequence of collisions and specific initial conditions imply that the discrete time evolution of the system is described by a discrete semigroup of natural powers of $\mathcal{E}$. May this this discrete set of channels be interpolated by a single one-parametric continuum of channels? Is this continuum a semigroup of channels, or not? Luckily, we have derived that the interpolating continuum for both cases of homogenization and decoherences. By deriving the corresponding master equations and testing the Markovianity of the generator we showed that these sets form one-parametric semigroups of channels. Let us note that a collision model simulation of given semigroups of channels is easy. Each semigroup $\mathcal{E}_t$ can be discretized by introducing a parameter τ and set $\mathcal{E}_n = \mathcal{E}_{n\tau}$. Since the choice of an interaction U inducing $\mathcal{E}_\tau$ is not unique, also the collision model is not unique. However, the particular collision dynamics of the system will be the same.

Let us note an interesting fact. Fix a unitary collision U. Even if ξ, ξ' induce a channels $\mathcal{E}, \mathcal{E}'$ with valid generators $\mathcal{G}, \mathcal{G}'$, the convex combination $\lambda\xi + (1 - \lambda)\xi'$ does not have to be associated with a correct Lindblad generator. There are collision models that cannot be interpolated by Markovian dynamics. Surprisingly, collision models provides richer dynamics than Lindblad's master equations. To sort out example consider Markovian models developed in order to describe true open system dynamics can be efficiently simulated (approximated) in collision models. We believe that such toy collision models provide an interesting playground capturing all the conceptual features of quantum open system dynamics.

Acknowledgments

We would like to thank coauthors of the original papers Peter Štelmachovič, Valerio Scarani, Nicolas Gisin, and Mark Hillery. Many thanks also to Daniel Burgarth for inspiring discussions on closely related topics, which would deserve some space in these lectures, but finally stayed outside the scope. We acknowledge financial support of the European Union projects HIP FP7-ICT-2007-C-221889, and CE QUTE ITMS NFP 26240120009, and of the projects, CE SAS QUTE, and MSM0021622419.

Bibliography

1. M. Ziman, P. Štelmachovič, V. Bužek, M. Hillery, V. Scarani, N. Gisin, Dilluting quantum information: An analysis of information transfer in system-reservoir interactions, Phys. Rev. A 65, 042105 (2002).
2. V. Scarani, M. Ziman, P. Štelmachovič, N. Gisin, V. Bužek, Thermalizing Quantum Machines: Dissipation and Entanglement, Phys. Rev. Lett. 88, 97905-1 (2002).
3. M. Ziman, P. Štelmachovič, V. Bužek, Description of quantum dynamics of open systems based on collision-like models, Open systems and information dynamics 12, No. 1, pp. 81-91 (2005).
4. M. Ziman, V. Bužek, All (qubit) decoherences: Complete characterization and physical implementations, Phys. Rev. A 72, 022110 (2005).
5. E. Schrödinger, An Undulatory Theory of the Mechanics of Atoms and Molecules, Phys. Rev. 28, 1049-1070 (1926).

6. E. B. Davies, *Quantum Theory of Open Systems* (Academic, London, 1976).
7. R. Alicki, K. Lendi, *Quantum Dynamical Semigroups and Applications*, Lecture Notes in Physics (Springer-Verlag, Berlin, 1987).
8. H. P. Breuer, *The Theory Of Open Quantum Systems* (Oxford University Press, USA, 2002).
9. M. A. Nielsen, I. L. Chuang, *Quantum Computation and Quantum Information* (Cambridge University Press, Cambridge, 2000).
10. T. Heinosaari, M. Ziman, Guide to mathematical concepts of quantum theory, Acta Physica Slovaca 58, 487-674 (2008).
11. V. Bužek and M. Hillery, Quantum copying: Beyond the no-cloning theorem, Phys. Rev. A 54, 1844 (1996).
12. V. Scarani, S. Iblisdir, N. Gisin, A. Acin, Quantum cloning, Rev. Mod. Phys. 77, 1225-1256 (2005).
13. M. Reed, B. Simon, *Methods of Modern Mathematical Physics I: Functional Analysis* (Academic Press, San Diego, 1980).
14. W. H. Zurek, Decoherence, einselection, and the quantum origins of the classical, Rev. Mod. Phys. **75**, 715 (2003).
15. F. Buscemi, G. Chiribella, G. M. D'Ariano, Inverting quantum decoherence by classical feedback from the environment, Phys. Rev. Lett. 95, 090501 (2005).
16. R. Horodecki, P. Horodecki, M. Horodecki, and K. Horodecki, Quantum entanglement, Rev. Mod. Phys. 81, pp. 865-942 (2009).
17. M. B. Plenio, S. Virmani, An introduction to entanglement measures, Quant. Inf. Comp. 7, 1 (2007).
18. W. K. Wootters, Entanglement of Formation of an Arbitrary State of Two Qubits, Phys. Rev. Lett. 80, 2245 (1998).
19. V. Coffman, J. Kundu, W. K. Wootters, Distributed Entanglement, Phys. Rev. A 61, 052306 (2000).
20. T. J. Osborne, F. Verstraete, General Monogamy Inequality for Bipartite Qubit Entanglement, Phys. Rev. Lett. 96, 220503 (2006).
21. H. Spohn, Kinetic equations from Hamiltonian dynamics: Markovian limit, Rev. Mod. Phys. 53, 569 (1980).
22. G. Lindblad, On the generators of quantum dynamical semigroups, Comm. Math. Phys. 48, pp. 119-130 (1976).
23. M. M. Wolf, J. I. Cirac: Dividing quantum channels, Comm. Math. Phys. 279, pp. 147-168 (2008).
24. L. V. Denisov, Infinitely divisible Markov mappings in the quantum theory of probability, Th. Prob. Appl. 33, 392 (1988).

AUTHOR INDEX